Crop Biotechnology

Genetic Modification and Genome Editing

Other Related Titles from World Scientific

An Introduction to Bioenergy
by Nigel G. Halford
ISBN: 978-1-78326-623-4
ISBN: 978-1-78326-624-1 (pbk)

Food Safety Assessment of Pesticide Residues
edited by Árpád Ambrus and Denis Hamilton
ISBN: 978-1-78634-168-6

Orchid Biotechnology III
edited by Wen-Huei Chen and Hong-Hwa Chen
ISBN: 978-981-3109-21-6

M. S. Swaminathan: Legend in Science and Beyond
by P. C. Kesavan
ISBN: 978-981-3200-09-8

Agroecological Practices for Sustainable Agriculture: Principles,
Applications, and Making the Transition
edited by Alexander Wezel
ISBN: 978-1-78634-305-5

Crop Biotechnology

Genetic Modification and Genome Editing

Nigel G Halford

Rothamsted Research, UK

World Scientific

NEW JERSEY · LONDON · SINGAPORE · BEIJING · SHANGHAI · HONG KONG · TAIPEI · CHENNAI · TOKYO

2019001452

Published by

World Scientific Publishing Europe Ltd.

57 Shelton Street, Covent Garden, London WC2H 9HE

Head office: 5 Toh Tuck Link, Singapore 596224

USA office: 27 Warren Street, Suite 401-402, Hackensack, NJ 07601

Library of Congress Cataloging-in-Publication Data
Names: Halford, Nigel G., author.
Title: Crop biotechnology : genetic modification and genome editing /
    by Nigel G. Halford (Rothamsted Research, UK).
Description: New Jersey : World Scientific, 2018.
Identifiers: LCCN 2017060976 | ISBN 9781786345301 (hc : alk. paper)
Subjects: LCSH: Plant biotechnology. | Plant genetics. | Transgenic plants.
Classification: LCC SB106.B56 H35 2018 | DDC 631.5/23--dc23
LC record available at https://lccn.loc.gov/2017060976

British Library Cataloguing-in-Publication Data
A catalogue record for this book is available from the British Library.

First published 2018 (Hardcover)
Reprinted 2019 (in paperback edition)
ISBN 978-1-78634-782-4 (pbk)

For any available supplementary material, please visit
https://www.worldscientific.com/worldscibooks/10.1142/Q0156#t=suppl

Desk Editors: Dr. Sree Meenakshi Sajani/Jennifer Brough/Koe Shi Ying

Typeset by Stallion Press
Email: enquiries@stallionpress.com

Preface

It is now more than two decades since the first large-scale cultivation of genetically modified (GM) crops and 15 years since the first edition of this book was published. Since then, the use of GM crops has continued to rise around the world, with 185 million hectares being grown in 2016. The number of traits introduced by genetic modification increases steadily, and although herbicide tolerance and insect resistance traits remain dominant, there have been exciting developments in the application of genetic modification to improve nutritional quality and food safety. There is still no established market for GM varieties of some of the world's major crops, including rice and wheat, but GM varieties of soybean, maize, cotton and oilseed rape, in particular, continue to be popular with farmers wherever they have been made available. Recent years have also seen further technological innovation in the development of a range of techniques that enable scientists to make specific changes to target genes. These techniques have been called genome editing, and the first genome edited commercial crop varieties are already in the field.

For crop scientists in the United Kingdom and the rest of Europe the situation remains as frustrating as when the first edition of this book was published in 2003. The GM crop debate of the late 1990s and early 2000s was lost and great damage was done to the European plant biotechnology industry as a result. The European regulatory system remains so difficult to negotiate that biotechnology companies are focussed on obtaining approval for their products to be imported into Europe so that they can be grown elsewhere rather than approval for their varieties to be cultivated in Europe. European farmers are therefore becoming increasingly disadvantaged as they have to compete with GM crop varieties that they are not allowed to grow. As non-GM crop products become more difficult to source it is recognised that the situation cannot continue as it is, but efforts to re-start the debate are greeted with the same shrill predictions of disaster and the same headlines as before. Turning that situation around will require strong leadership and there is no sign at all of where that leadership might come from. Meanwhile, countries across the Americas and Asia, in particular, continue to extend their lead in developing what will be a key 21st Century technology.

This edition of the book is expanded and updated to cover the latest developments in the techniques and applications of crop biotechnology and the legislation that controls its use, and all sections are extended to include genome editing as well as genetic modification. The motivation for writing the book continues to be to provide answers to legitimate questions about crop biotechnology: what is it, how is it done, how does it differ from what has been done before, is it safe, how is it regulated and what implications does it have for plant breeding, agriculture and the environment?

Nigel G. Halford

About the Author

 Professor Nigel G. Halford is a Research Leader at Rothamsted Research, the UK's largest and oldest crop and agricultural research centre. He obtained his first degree from the University of Liverpool in 1983 and a Masters degree from University College London in 1984. He studied the genes encoding a family of wheat seed proteins for his PhD while at Rothamsted in the 1980s, then spent 11 years at Long Ashton Research Station near Bristol before returning to Rothamsted in 2002. He is the author of more than 140 scientific papers, and as well as this title and its earlier editions entitled 'Genetically Modified Crops', has written 'An Introduction to Bioenergy' and edited books on 'Protein Phosphorylation in Plants', 'Plant Biotechnology', and 'Energy Crops'. His research programme concerns the genetics of metabolic regulation in crop plants, how plant metabolism is affected by environmental stress and crop management, and how it can be manipulated to improve crop yield, quality and food safety. Professor Halford has worked with GM plants for nearly 30 years and has been a prominent participant in the GM debate since the 1990s. He is a Visiting Professor at Shanghai Academy of Agricultural Sciences and was awarded

The Magnolia Silver Medal by the Shanghai Municipal People's Government in 2012. He is also Special Professor at the University of Nottingham, a former member of the UK's Advisory Committee for Animal Feedingstuffs (ACAF), and a Fellow of the Royal Society of Biology.

Contents

Chapter 1

DNA, Genes, Genomes and Plant Breeding

1.1. A Brief History of Genetics

To many people, the beginnings of genetics can be traced back to the publication in 1859 of Charles Darwin's book, 'On the Origin of Species by Means of Natural Selection'. This established the theory of evolution based on the principle of natural selection, discovered independently at the same time by Alfred Russell Wallace, and was the culmination of decades spent collecting and examining evidence.

Darwin argued that species and individuals within species compete with each other. Furthermore, individuals within a species are not all the same; they differ, or show variation. Those that are best fitted for their environment are the most likely to survive, reproduce and pass on their characteristics to the next generation. If the environment changes or a species colonises a new environment, different characteristics may be selected for, leading to change or evolution. The diversity of life on Earth could therefore be explained by the adaptation of species and groups of individuals within species to different and changing environments, leading to the extinction of some species and the appearance of others. Species that were similar had arisen from a recent common ancestor. Most controversially, humans were similar to other apes not because God had made it so but because humans and other apes had a relatively recent common ancestor.

Darwin was not the only scientist whose thinking was challenging the notion of Life's diversity arising from supernatural creation. Georges Cuvier, in Paris, had noted that fossils in deep rock strata were less like living animals than those in shallow strata. At the same time, studies were revealing that the anatomies of different animals were based on the same internal patterns, and Jean-Baptiste Lamarck had already published his theory that animals could transform into one another.

There was a problem with the theory of evolution by natural selection, and Darwin was well aware of it. At that time, the traits of parents were believed to be mixed in their offspring so that the characteristics of the offspring would always be intermediate between those of the parents. If this were true, natural selection as Darwin proposed it could not work because the process requires there to be variation within a population so that differences can be selected for. Blending traits would have the effect of reducing variation with every successive generation. Darwin even considered variations on Lamarck's theory that changes acquired during an organism's lifetime could be passed on to its offspring.

The solution of the problem was found by Gregor Mendel but, ironically, Darwin was never aware of Mendel's work. In 1857, Mendel performed some experiments with pea plants in the garden of the Austrian monastery where he was a monk. Mendel recorded the different characteristics of the plants, such as height, seed colour, seed coat colour and pod shape, and observed that offspring sometimes, but not always, showed these same characteristics. In his first experiments, he self-pollinated short and tall plants and found that they bred true, the short having short offspring and the tall having tall offspring. However, when he crossed short and tall plants he found that all of the offspring (the F1 generation) were tall. He crossed the offspring again and the short characteristic reappeared in about a quarter of the next generation (the F2 generation).

Many years and experiments later, Mendel concluded that characteristics were passed from one generation to the next in pairs, one from each parent, and that some characteristics were dominant over others. His most famous experiments concerning the inheritance of a trait in which the seeds were wrinkled are still taught in schools today. His findings were published but ignored for decades as the work of an amateur. Later, they became known as the Mendelian Laws and can be regarded as the foundation of modern genetics and plant breeding.

The significance of Mendel's work was that it showed that whether or not the offspring of two parents resemble one parent or are an intermediate between the two, they inherit a single unit of inheritance from each parent. These units are reshuffled in every generation and traits can reappear, so variation is not lost. Units of inheritance subsequently became known as genes.

The next big advance came in 1902, when a British doctor, Sir Archibald Garrod, studied an inherited human disease, alkaptonuria. Alkaptonuria sufferers excrete dark red urine because they lack an enzyme that breaks down the reddening agent, alkapton. Garrod noted that alkaptonuria recurred in families and that parents of sufferers were often closely related to each other. In other words, it was an inherited condition, and Garrod referred to alkaptonuria and similar diseases as 'inborn errors of metabolism'. The significance of Garrod's work was that it made the link between the inheritance of one particular gene and the activity of a single protein.

Like Mendel, Garrod was ahead of his time and his work was forgotten until the link between genes and proteins was made again in the 1930s by American geneticists George Beadle and Edward Tatum. Beadle and Tatum showed that a mutation in the fungus, *Neurospora crassa*, affected the synthesis of a single enzyme required to make an essential nutrient, and that the mutation was inherited through successive generations. This led to Beadle and Tatum publishing the one gene, one enzyme hypothesis in 1941. The hypothesis still stands today, although it has been modified slightly to account for the fact that some proteins are made up of more than one subunit and the subunits may be encoded by different genes.

1.2. Deoxyribonucleic Acid (DNA)

With the principles of inheritance established, the search was on for the substance through which the instructions for life were passed from one generation to the next. The conclusive experiment that identified this substance is now accepted to be that conducted by Oswald Avery, Colin

MacLeod and Maclyn McCarty at the Rockefeller Institute Hospital in 1944. The experiment showed that the transfer of a DNA molecule from one strain of a bacterium, *Streptomyces pneumoniae*, to another changed its virulence. This showed that DNA was the substance that contained the information for an organism's development: the genetic material.

DNA had been discovered in 1869 by Friedrich Miescher, who worked at the Physiological Laboratory of the University of Basel and in Tübingen. All plants, animals, fungi and bacteria contain DNA, and they pass a copy of their DNA to their offspring. The evolution of life on Earth was wholly dependent on the extraordinary nature of DNA, and the discovery and characterisation of DNA and how it functions as the genetic material is one of the great achievements of science.

DNA consists of a backbone of units based on deoxyribose, a type of sugar, linked by phosphate groups (Figure 1.1). In theory, there is no limit to the length of the deoxyribose chain and DNA molecules can be extremely large. Attached to each deoxyribose in the chain is an organic base, of which there are four kinds: adenine, cytosine, guanine and thymine (Figure 1.1). Any base can be present at any position, and it is this that gives DNA its variability. A unit of deoxyribose, a phosphate group and an organic base is called a nucleotide, with the different nucleotides often referred to in shorthand as A, C, G or T, according to the organic base, adenine, cytosine, guanine or thymine, that they contain. Information is encoded within DNA as the sequence of nucleotides in the chain. In other words, all of the instructions for life on Earth are written in a language of four letters, A, C, G and T.

The deduction of the 3D structure of DNA is famously attributed to James Watson and Francis Crick, who worked together in Cambridge and published the structure in 1953. In fact, the breakthrough owed as much to the work of Maurice Wilkins and Rosalind Franklin, two scientists from a team led by William Bragg, who were pioneering the techniques of X-ray crystallography in the study of large molecules in Cambridge at that time. Watson and Crick were inspired by the work of Linus Pauling, who had discovered that the molecules of some proteins have helical shapes. They designed several models of DNA and finally came up with the correct structure after analysing Franklin's X-ray photographs. Watson, Crick and Wilkins jointly received the Nobel Prize for Physiology

(a)

(b)

Figure 1.1. Structure of DNA. (a) A single strand of DNA has a backbone of sugars (deoxyribose) linked by phosphate groups. One of four organic bases, adenine, cytosine, guanine or thymine, is attached to each deoxyribose molecule. (b) Double-stranded DNA consists of two strands running in opposite directions, held together by hydrogen bonds between the bases. The bases can occur in any order but they must be complementary: adenine on one strand is always paired with thymine on the other, while cytosine is always paired with guanine.

or Medicine in 1962. Tragically, Rosalind Franklin died of cancer in 1958 and could not be honoured because the Nobel Prize is not awarded posthumously.

The structure of DNA, as deduced by Watson and Crick, consists of two deoxyribose chains running in opposite directions coiled around each other to form a beautiful structure called a double helix (Figure 1.2). The bases attached to each deoxyribose are on the inside of the helix at right angles to the helix axis, and the structure is stabilised by hydrogen bonds between adjacent bases from each chain (Figures 1.1 and 1.2). Because of the dimensions of the helix and the structures of the bases, adenine on one chain must always be faced with thymine on the other, while guanine is always paired with cytosine. These pairs, A and T or C and G, are often referred to as base pairs (bp), and the size or length of a DNA molecule is often given in base pairs.

This specific pairing of the bases is important because it underlies the process by which DNA can be replicated as an exact copy. If double-stranded DNA is unravelled to form two single strands, each strand can

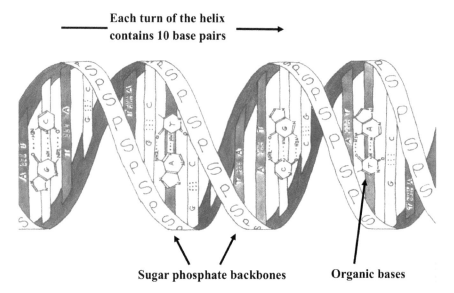

Each turn of the helix contains 10 base pairs

Sugar phosphate backbones   **Organic bases**

Figure 1.2.   Structure of a DNA double helix. S = sugar, P = phosphate. The organic bases, adenine, cytosine, guanine and thymine, are labelled A, C, G and T. Drawn by Sandra Hey.

act as a template for the synthesis of a complementary chain, and two replicas of the original double-stranded molecule are created.

1.3. Genes

We can now define genes, the units of heredity, as functional units within a DNA molecule. A gene will contain the information not only for the structure of a protein, but also for when and where in an organism the gene is active. This information is encoded in the sequence of base pairs within the region of the DNA molecule that makes up the gene. Genes can comprise over a million base pairs but are usually much smaller, averaging 3,000 bp.

The part of the gene containing the information for the primary structure of the protein encoded by the gene is called the coding region (Figure 1.3). The sequence of amino acids in the protein is determined by the sequence of base pairs in the coding region of the gene, each amino acid in the protein being represented by a triplet of base pairs (a codon) in the DNA sequence. Adjacent to and usually (but not always) 'upstream' of the coding sequence are the regulatory regions of the gene that determine when and where the gene is active (or expressed; Section 1.4). The region containing this information is called the gene promoter. The region 'downstream' of the coding region is called the gene terminator.

The regulatory information in the gene promoter is in the form of short nucleotide sequences that are recognised and bound by proteins called transcription factors. These sequences are called promoter elements or regulatory motifs and can have positive effects (enhancers) or negative effects (silencers) on the activity of the gene.

1.4. Gene Expression

The term gene expression refers to the process by which the protein that is encoded by the gene is produced. It comprises the following sequence of events:

1. The DNA molecule is used as a template for the synthesis of a related molecule called ribonucleic acid (RNA). RNA, like DNA, comprises

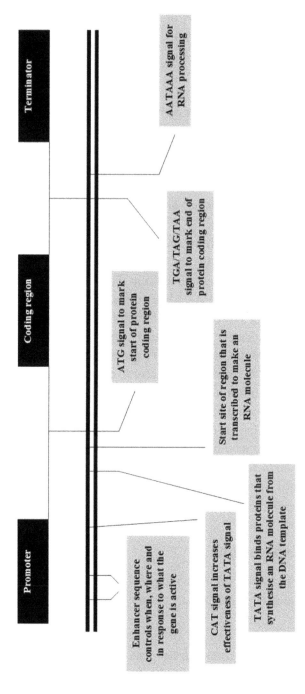

Figure 1.3.   Schematic diagram of the structure of a gene.

organic bases on a sugar-phosphate backbone, but it differs from DNA in that it is generally single-stranded and relatively short, the deoxyribose of DNA is replaced with a similar sugar, called ribose, and while three of the four organic bases in DNA are also present in RNA, thymine is replaced with a different organic base called uracil. Nevertheless, the sequence of bases on the newly synthesised RNA molecule is determined by the sequence of bases on the DNA template; hence, the encoded information is passed from the DNA to the RNA molecule. This process is called transcription.

2. The RNA molecule is processed and transported to protein complexes called ribosomes where protein synthesis occurs.

3. A protein is synthesised, the amino acid sequence of which is specified by the RNA molecule. This process is called translation.

At the end of this process, information encoded in the DNA molecule has been transferred through an RNA molecule to the protein production machinery and used to make a protein.

Genes that are active (i.e. being expressed) throughout an organism all of the time are referred to as constitutive or housekeeping genes. Most genes are not constitutive, but are subject to various kinds of regulation. Some are expressed only in certain organs, tissues or cell types. Others are expressed during specific developmental stages of an organism, or in response to a stimulus. In the case of plants, gene expression responds to a host of stimuli, including light, temperature, freezing, grazing, disease, shading and nutritional status. This regulation is brought about by interactions between promoter elements and transcription factors. DNA is, therefore, not just the blueprint for an organism's development; it is active throughout an organism's life and plays a central role in an organism's interactions with its environment.

1.5. Genomes

The genome is defined as an organism's complete complement of DNA, and the study of genomes is called genomics. Genomes can be very large, but it is now possible to determine the sequence of base pairs for entire genomes from even complex animals, plants and fungi. The entire yeast genome base pair sequence was determined in 1996, and a draft of the human genome sequence was completed in 2000, with a 'finished' sequence produced by 2003. The base pair sequences of entire genomes have also been determined for *Arabidopsis* (*Arabidopsis thaliana*) (a plant also known as thale cress that is used as a 'model' in plant genetics because of its relatively small genome), Brachypodium (*Brachypodium distachyon*) (a model grass), rice, soybean, lotus, poplar, sorghum, wine grape, maize, wheat and potato.

The human genome contains approximately 3 billion bp, organized into 23 chromosomes ranging from 50 to 250 million bp long. The rice genome contains 466 million bp, while that of *Arabidopsis* contains approximately 126 million bp and that of wheat a daunting 16 billion bp. The human genome, however, contains only 30–40,000 genes, while that of rice contains 45–56,000 genes, with wheat containing a similar number (it could be argued that rice and wheat are more complex and highly evolved organisms than humans). The reason for the discrepancy between genome size and gene number is that much of the genome does not contain genes. In fact, genes represent only 2% of the human genome. The rest of the DNA is sometimes referred to as 'junk DNA', and the amount of it varies greatly between different species. Much of it is highly repetitive in its base pair sequence and may promote rearrangement of the DNA molecules, driving the formation of new genes and therefore variation and evolution.

The technology for determining the sequence of base pairs in DNA is still developing rapidly, and so-called 'next-generation sequencing' (NGS) technologies are enabling DNA sequence information to be generated at a throughput and price that was not dreamt of at the turn of the century. The Human Genome Project was approved by the US Government in 1990. A draft genome sequence was published in 1999 and a finished sequence in 2003. The total cost of the project was slightly under the predicted budget at $2.7 billion, or around $1 per bp, whereas the current cost of DNA sequence analysis is under $1 per megabase (million base pairs). The process has also become much quicker: Entire genome sequences can now be obtained in weeks or months rather than years, and the time can be shortened further by the process of exon capture, selecting only the parts of the genome that encode proteins.

1.6. Genetic Change

Genetic change is a natural and desirable process. It results in variation in shape, form and behaviour of the individuals within a species, allowing for evolution and adaptation. It is crucial for the survival of any species and the evolution of new species in response to environmental change.

The process of genetic change in nature is called evolution, and when driven by natural selection it results in adaptive improvement. In the absence of selection, it leads to diversity and variation, making it more likely that some individuals will survive if the environment changes.

1.7. Plant Breeding

Natural selection is not the only driver for genetic change at work in the world today. Man has discovered that rapid genetic change can be induced by the artificial selection of individuals for breeding. This has been used to produce the Chihuahua, the greyhound and the bull terrier from the wolf, thoroughbred racehorses, the Shire horse and the Shetland pony, and every modern farm animal. It has also been applied over millennia for crop improvement (Figure 1.4).

It is probable that crop improvement has been practised since humankind first started to plant and harvest crops, rather than forage for food from wild plants. This is thought to have become widespread about 10,000 years ago. At first, such improvement may well have occurred unconsciously, through the selection of the most vigorous individuals from highly variable populations for planting in the following year, but then became more systematic. Types of a crop plant with different characteristics would be grown in adjacent plots and some of the seed produced would result from crossing of the two types. Farmers would then select the best seed for the next generation. This relatively primitive but effective form of plant breeding is still used in many parts of the world today, and through the ages has changed crop plants greatly from their wild ancestors and relatives.

This is illustrated quite dramatically in Figure 1.5, in which grain from a modern breadmaking wheat variety is compared with grain from two wild relatives. Grain that was buried in tombs in ancient Egypt and

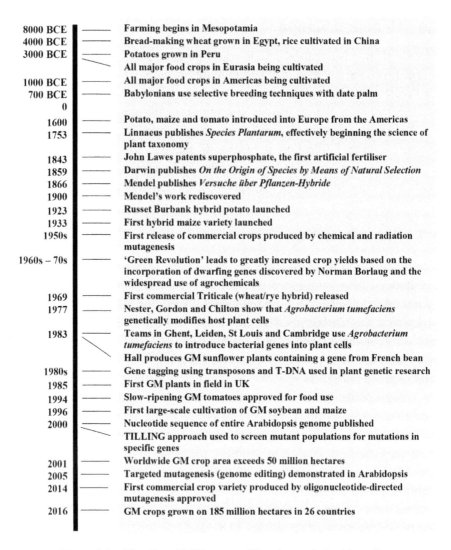

8000 BCE	Farming begins in Mesopotamia
4000 BCE	Bread-making wheat grown in Egypt, rice cultivated in China
3000 BCE	Potatoes grown in Peru
	All major food crops in Eurasia being cultivated
1000 BCE	All major food crops in Americas being cultivated
700 BCE	Babylonians use selective breeding techniques with date palm
0	
1600	Potato, maize and tomato introduced into Europe from the Americas
1753	Linnaeus publishes *Species Plantarum*, effectively beginning the science of plant taxonomy
1843	John Lawes patents superphosphate, the first artificial fertiliser
1859	Darwin publishes *On the Origin of Species by Means of Natural Selection*
1866	Mendel publishes *Versuche über Pflanzen-Hybride*
1900	Mendel's work rediscovered
1923	Russet Burbank hybrid potato launched
1933	First hybrid maize variety launched
1950s	First release of commercial crops produced by chemical and radiation mutagenesis
1960s – 70s	'Green Revolution' leads to greatly increased crop yields based on the incorporation of dwarfing genes discovered by Norman Borlaug and the widespread use of agrochemicals
1969	First commercial Triticale (wheat/rye hybrid) released
1977	Nester, Gordon and Chilton show that *Agrobacterium tumefaciens* genetically modifies host plant cells
1983	Teams in Ghent, Leiden, St Louis and Cambridge use *Agrobacterium tumefaciens* to introduce bacterial genes into plant cells
	Hall produces GM sunflower plants containing a gene from French bean
1980s	Gene tagging using transposons and T-DNA used in plant genetic research
1985	First GM plants in field in UK
1994	Slow-ripening GM tomatoes approved for food use
1996	First large-scale cultivation of GM soybean and maize
2000	Nucleotide sequence of entire Arabidopsis genome published
	TILLING approach used to screen mutant populations for mutations in specific genes
2001	Worldwide GM crop area exceeds 50 million hectares
2005	Targeted mutagenesis (genome editing) demonstrated in Arabidopsis
2014	First commercial crop variety produced by oligonucleotide-directed mutagenesis approved
2016	GM crops grown on 185 million hectares in 26 countries

Figure 1.4.   Timeline: 10,000 years of farming and plant breeding.

discovered in modern times when the tombs were opened was found to be much more similar to modern wheat grain than to grain of wild wheat. So, even by that time, 4,000 or more years ago, wheat had been changed greatly by simple selection. Wheat cultivation and improvement enabled the production of food in large quantities, and it is doubtful whether

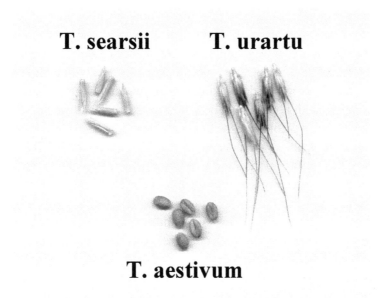

Figure 1.5. Grain of bread wheat (*Triticum aestivum*) compared with that of two wild relatives, *Triticum searsii* and *Triticum urartu*. Bread wheat is a product of thousands of years of selective breeding.

modern western civilisation would have developed without it; a similar argument could be made for the dependence of eastern civilisation on rice. The development of modern wheat is a good example of how the genetics of crop plants have been manipulated by farmers and breeders who for millennia knew nothing about the molecular basis of what they were doing.

Wheat actually comprises many different species, some with one, two or even three genomes (diploids, tetraploids and hexaploids, respectively). The tetraploids, which include cultivated pasta wheats, arose through hybridisation and genome duplication between ancestral diploid species. Breadmaking wheat is a hexaploid that arose through hybridisation between a tetraploid and a third diploid species. There are no wild hexaploid wheat species: breadmaking wheat appeared within cultivation in south-west Asia approximately 10,000 years ago, and its use spread westwards into Europe.

Another good example of the remarkable changes that have been introduced into crop plants by simple selection over many generations is the cabbage family of vegetables, which includes kale, cabbage, cauliflower, broccoli and Brussels sprouts (Figure 1.6). The wild relative of the cabbage family grows in the Mediterranean region of Europe, and it was first domesticated approximately 7,000 years ago. Through selective breeding, the crop plants became larger and leafier, until a plant very similar to modern kale was produced in the 5th Century BCE. By the 1st Century CE, a different variation was being grown alongside kale. It had a cluster of tender young leaves at the top of the plant and is known today as cabbage.

In the 15th Century, cauliflower was produced in Southern Europe by selecting plants with large, tender, edible flowering heads, and broccoli was produced in similar fashion in Italy about a century later. The last variation to appear was Brussels sprouts, which were bred in Belgium in

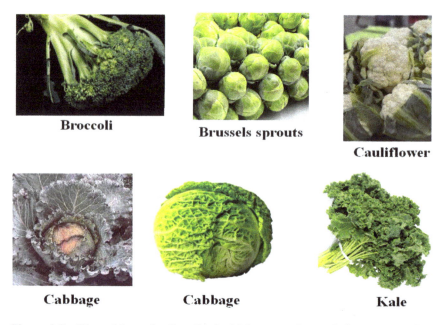

Broccoli

Brussels sprouts

Cauliflower

Cabbage

Cabbage

Kale

Figure 1.6. The cabbage family, all of which are variants of the same species: *B. oleracea.*

the 18th Century, with large buds along the stem. All of these very different vegetables are actually variants of the same species, *Brassica oleracea*.

1.8. Modern Plant Breeding

True scientific breeding dates from about 1900 (Figure 1.4) when the re-discovery of the work of Gregor Mendel on the inheritance of characteristics in pea provided a sound theoretical basis (Section 1.1). As described in Section 1.5, plants contain many genes, estimated at about 26,000 in *Arabidopsis* and 45,000–56,000 in rice. These genes may exist in different forms (called alleles) in individuals of the same species, in the same way, for example, that alleles of the gene that imparts eye colour exist in humans. Crossing of individuals with contrasting characteristics results in a population of individuals with different combinations of alleles from the two parents. At least some of these combinations may result in improved performance relative to either parent, and the identification and further improvement of these is the task of the plant breeder. Since the breeder is literally dealing with thousands of genes, the task is formidable.

Nevertheless, plant breeders have been very effective, and it is just as well that they have. At the end of the 18th Century, Reverend Thomas Malthus predicted in his 'Essay on the Principle of Population', which he published anonymously, that food supply could not keep up with exponentially rising population growth. At the time, world population was approximately one billion. In 1999, the world population reached six billion, and it passed seven billion in 2011. *Per capita* consumption has also risen in many parts of the world: in China, for instance, annual *per capita* consumption of wheat rose from 28 kg in 1962 to 92 kg in 2012, a more than threefold increase in half a century. And yet, with the exception of sub-Saharan Africa, the world has generally not been critically short of food, because food demand has been met through dramatic increases in crop yield. As an example, the yield per hectare of wheat grown in the United Kingdom from 1961 to 2014 is shown in Figure 1.7. Yield approximately doubled between the 1960s and the 1990s, before levelling off at around 8 tonnes per hectare, while production approximately quadrupled from less than 4 to around 15–16 million tonnes. Of course, plant breeding was not the only contributor to this increase.

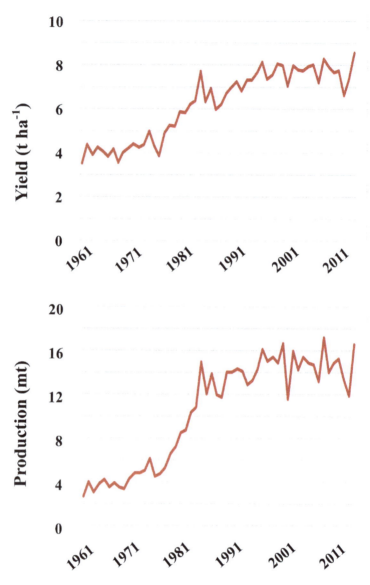

Figure 1.7.   **United Kingdom wheat yield (tonnes per hectare) and production (million tonnes) from 1961 to 2014. Data from United Nations Food and Agriculture Organisation.**

Mechanisation, the development and use of nitrogen fertilizers, herbicides and pesticides and other improvements in farm practice played a part. Indeed, it was the combination of mechanisation, the widespread use of agrochemicals and the incorporation of dwarfing genes into cereals that led to the 'Green Revolution' of the 1960s and 1970s. The dwarfing genes concerned actually affected the synthesis of a plant hormone, gibberellin, although that was not known at the time. Their incorporation reduced the amount of resources that crop plants put into their inedible parts and at the same time made crops less susceptible to lodging (falling over in damp and/or windy conditions).

Plant breeding has also been successful in improving food safety. Contrary to the popular notion that natural is 'good' and man-made is 'bad' when it comes to food, plants did not evolve to be eaten. In fact, with the exception of fruit, the opposite is true, and since plants are unable to flee a large animal or brush off an insect and do not have an immune system like that of animals to fight off microorganisms, they have evolved a sophisticated armoury of chemical weapons with which to defend themselves. Many of these chemicals are potentially toxic or allergenic to humans and they are most abundant in seeds and tubers, whose rich reserves of proteins, starch and oil are particularly attractive to herbivores, pests and pathogens. Well-known examples are glycoalkaloids in potatoes, cyanogenic glycosides in linseed, proteinase inhibitors in soybean and other legume seeds and glucosinolates in *Brassica* oilseeds.

Plant breeders have made great strides in reducing the levels of these chemicals in crop plants, although some crops have only been regarded as fit for human consumption for a surprisingly short time. Oilseed rape, for example, was first grown in the UK during World War II to provide oil for industrial uses and some varieties are still grown for that purpose. Its oil was regarded as unfit for human consumption because it contained high levels of erucic acid and glucosinolates. Erucic acid (Section 3.8) is a fatty acid that has been linked with the formation of fatty deposits in heart muscle and consequent muscle damage. Glucosinolates have a bitter/hot flavour and are poisonous, although their toxicity is complicated by the fact that they break down in the body to a variety of different compounds, with different glucosinolates producing different breakdown products.

It has been suggested that they are responsible for oilseed rape having relatively few specialised pests. Glucosinolates are present at even higher levels in the meal that remains after oil has been extracted, with implications for the use of the meal in animal feed. After World War II, breeders gradually reduced the levels of erucic acid and glucosinolates, but the oil was only passed fit for human consumption and the meal for animal feed in the 1980s.

1.8.1. *Wide and forced crossing and embryo rescue*

There is a major limitation to the success of plant breeding, which is the extent of variation in the parental lines. Plant breeders cannot select for variation that is not present in their breeding population. They have therefore looked further afield, to exotic varieties and wild relatives of crop species, for genes to confer important characters. This 'wide crossing' would not occur in nature and usually requires 'rescue' of the embryo to prevent abortion. This entails surface-sterilizing developing seeds and dissecting them under a microscope to remove the embryos. The embryos are then cultured in a nutrient medium until they germinate and develop into seedlings, which are then transferred to soil.

Forced crosses of this sort may be carried out to transfer genes for single characters (disease resistance, for example). However, tens of thousands of other genes will inevitably be brought into the breeding programme alongside the desired gene. Most of these can be eliminated by crossing the hybrid plant repeatedly with the crop parent (back-crossing). Nevertheless, many genes of unknown function will remain at the end of this process, and if they have undesirable effects on the performance of the crop, the programme will fail.

Forced crosses have even been made between different plant species. The best-known example of this is triticale, a hybrid between wheat and rye. The first deliberately made hybrid between these species was reported by A. S. Wilson in Scotland in 1875. However, these plants were sterile because the chromosomes from wheat and rye will not form pairs during meiosis. This can be overcome by inducing chromosome doubling by chemical treatment, usually with colchicine. The cross is usually made between durum wheat, a tetraploid, and rye to produce a hexaploid triticale, although it is also possible to cross hexaploid wheat with rye to produce an octoploid triticale.

The name triticale was used first in 1935 by Tschermak, but it was not until 1969, after considerable improvement through breeding, that the first commercial varieties of triticale were released. Triticale is now grown on more than 2.4 million hectares worldwide, producing more than 6 million tonnes of grain per year. Its advantage is that it possesses the yield potential of wheat and the hardiness (tolerance of acid soils, damp conditions and extreme temperatures) of rye. However, it does not match the bread-making quality of wheat, and so, up to now, it has been used mostly for animal feed. It may have potential as a raw material for bioethanol production (Section 3.10).

1.8.2. *Radiation and chemical mutagenesis*

Another way of increasing the variation within a breeding population is to introduce mutations artificially. This is usually done with seeds and involves treatment with neutrons, gamma rays, X-rays, UV radiation or a chemical (a mutagen). All of these treatments damage DNA. If there is too much damage the seed will die, but minor damage can be repaired, sometimes resulting in changes in the DNA sequence. Even small changes of this sort may be lethal, but occasionally changes are made to a gene that leave the seed viable but alter the characteristics of the plant. The process is entirely random, and mutagenesis programmes usually involve very large populations of at least 10,000 individuals.

The first attempts to produce plant mutants were made in the 1920s and the first commercial varieties arising from mutation breeding programmes became available in the 1950s. The technique was particularly fashionable in the 1960s and 1970s but continues to be used today. One of the earliest cultivars produced by mutagenesis was the oilseed rape cultivar Regina II, which was released in Canada in 1953. Mutagenesis played an important role in the improvement of oil quality of oilseed rape and flax, and in durum wheat breeding in Italy. There are believed to be hundreds of rice varieties that incorporate artificially induced mutations, and most North American white bean varieties incorporate a mutation that was first induced by X-ray treatment.

The best known example of an irradiation mutant grown in the UK is the barley variety, Golden Promise. This was the most successful UK malting barley variety in the latter part of the last century, and although it is not grown widely now it has been incorporated into many barley breeding programmes.

1.9. The Advent of Genetic Modification and Genome Editing

Crossing different varieties and species involves the mixing of tens of thousands of genes, sometimes with unpredictable results, while chemical and radiation mutagenesis are, of course, random processes and it is impossible to count, let alone characterize, all the effects of a mutagenesis programme. Radiation and chemical mutants are likely to carry a baggage of uncharacterised genetic changes with unknown effects. Nevertheless, these techniques were developed and applied to plant breeding at a time when food production was a much higher priority than food safety and the risks were not a matter of public debate. They have now been around for so long that they are considered to be part of 'traditional' plant breeding, and crops modified by wide crossing and mutation breeding are accepted readily.

Even with the possibility of crossing crop plants with exotic relatives and related species and introducing mutations artificially, the variation available to plant breeders remained limited. Both techniques also had the disadvantage of introducing unwanted as well as desirable genetic changes. In the late 1970s, a new technique became available, that of inserting specific genes into the genome of a plant artificially. It meant that, technically, there was no limit to the source of new genes and it enabled plant breeders to bring specific genes into breeding programmes without unwanted genetic baggage. This technique was first called genetic engineering, and subsequently genetic modification.

In the first edition of this book, I referred to genetic modification as a new technique, but it is now inappropriate to do so. Genetically modified (GM) crops have been grown commercially for a generation. Furthermore, genetic modification did not represent the last innovation in plant breeding: recent years have seen the use of genomics to speed up 'conventional' plant breeding, and the development and application of a range of techniques that enable scientists to make specific changes to target genes (in other words they can be viewed as targeted mutagenesis). There are a variety of them, but they have been given the umbrella term genome (or gene) editing. Some involve a genetic modification step, but even when that is the case, they can usually result in a plant in which the genetic modification is no longer present but a native gene has been 'edited'. The next chapter describes the techniques used in genetic modification and genome editing.

Chapter 2

The Techniques of
Plant Genetic Modification
and Genome Editing

Any new crop variety has been changed genetically, using the methods described in the previous chapter. However, the term genetically modified (GM) is a relatively new expression that describes a plant that contains a gene or genes that have been introduced artificially. GM plants can also be described as being transgenic or having been transformed, or as containing 'foreign' DNA (DNA from a different organism). The term genetically engineered (GE) can be used instead of GM.

Plant genetic modification became possible in the late 1970s as a result of the development of techniques for manipulating DNA in the laboratory and introducing it into the DNA of a plant. Then from the 1990s onwards, a series of techniques were developed to enable the targeted mutagenesis of plant genomes; in other words, the introduction of single-nucleotide changes, or the deletion or insertion of short stretches of DNA, at targeted positions in a plant's genome. These techniques, some of which involve a genetic modification step, have been given the umbrella term genome or gene editing. Even those that involve a genetic modification step can be used to generate plants that contain no foreign DNA, so the plants produced at the end of the process are generally regarded as not being GM plants.

2.1. A Brief History of the Development of Recombinant DNA Technology

In the quarter century after the description of the structure of DNA by Watson and Crick in 1953, huge strides were made in the study of DNA and the enzymes present in cells that can work on it (Figure 2.1). In 1955, Arthur Kornberg at Stanford University isolated DNA polymerase, an enzyme that synthesises DNA. He received the Nobel Prize for Medicine in 1959. In 1966, Bernard Weiss and Charles Richardson, working at Johns Hopkins University, isolated DNA ligase, an enzyme that 'glues' two ends of DNA together. In 1970, Hamilton Smith, also at Johns Hopkins University, described the first characterisation of a restriction endonuclease (now usually called a restriction enzyme), an enzyme that had the ability to recognise specific short sequences of base pairs in a DNA molecule and cut the molecule at that point. Smith shared the Nobel Prize for Medicine in 1978.

The rapid progress continued. In 1972, Paul Berg at Stanford reported that he had constructed a DNA molecule by cutting viral and bacterial DNA molecules with restriction enzymes and then recombining them. He received the Nobel Prize for Chemistry in 1980. Then in 1973, Stanley Cohen and Annie Chang of Stanford together with Herbert Boyer and Robert Helling of the University of California, San Francisco, demon-strated that DNA that had been cut with a restriction enzyme could be recombined with small, self-replicating DNA molecules from bacteria, called plasmids. The product, a recombinant plasmid, could then be rein-troduced into bacterial cells and would replicate. In 1975, a British scien-tist, Edwin Southern, working at the University of Edinburgh, reported the transfer of DNA fragments that had been separated by electrophoresis to a membrane and the detection of specific DNA fragments by hybridisation with DNA probes. This technique became known as Southern blotting. Then, in 1977, Walter Gilbert at Harvard and Fred Sanger in Cambridge separately developed methods for determining the sequence of nucleo-tides in a DNA molecule. Sanger and Gilbert shared the Nobel Prize for Chemistry with Berg in 1980.

Scientists now had the tools for cutting DNA molecules at specific points and gluing them back together in different combinations to make

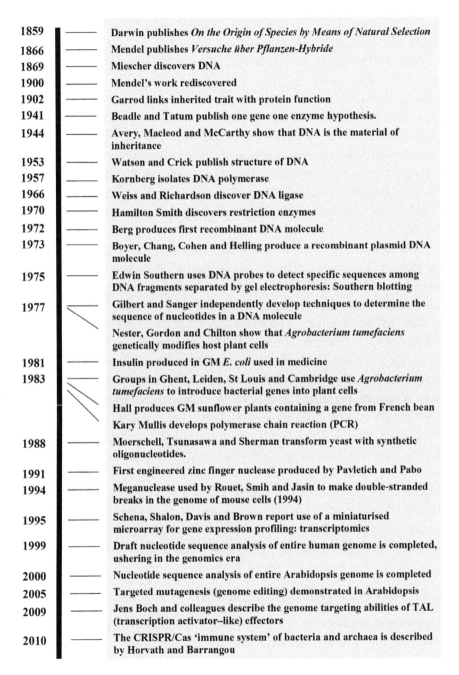

1859	Darwin publishes *On the Origin of Species by Means of Natural Selection*
1866	Mendel publishes *Versuche über Pflanzen-Hybride*
1869	Miescher discovers DNA
1900	Mendel's work rediscovered
1902	Garrod links inherited trait with protein function
1941	Beadle and Tatum publish one gene one enzyme hypothesis.
1944	Avery, Macleod and McCarthy show that DNA is the material of inheritance
1953	Watson and Crick publish structure of DNA
1957	Kornberg isolates DNA polymerase
1966	Weiss and Richardson discover DNA ligase
1970	Hamilton Smith discovers restriction enzymes
1972	Berg produces first recombinant DNA molecule
1973	Boyer, Chang, Cohen and Helling produce a recombinant plasmid DNA molecule
1975	Edwin Southern uses DNA probes to detect specific sequences among DNA fragments separated by gel electrophoresis: Southern blotting
1977	Gilbert and Sanger independently develop techniques to determine the sequence of nucleotides in a DNA molecule
	Nester, Gordon and Chilton show that *Agrobacterium tumefaciens* genetically modifies host plant cells
1981	Insulin produced in GM *E. coli* used in medicine
1983	Groups in Ghent, Leiden, St Louis and Cambridge use *Agrobacterium tumefaciens* to introduce bacterial genes into plant cells
	Hall produces GM sunflower plants containing a gene from French bean
	Kary Mullis develops polymerase chain reaction (PCR)
1988	Moerschell, Tsunasawa and Sherman transform yeast with synthetic oligonucleotides.
1991	First engineered zinc finger nuclease produced by Pavletich and Pabo
1994	Meganuclease used by Rouet, Smih and Jasin to make double-stranded breaks in the genome of mouse cells (1994)
1995	Schena, Shalon, Davis and Brown report use of a miniaturised microarray for gene expression profiling: transcriptomics
1999	Draft nucleotide sequence analysis of entire human genome is completed, ushering in the genomics era
2000	Nucleotide sequence analysis of entire Arabidopsis genome is completed
2005	Targeted mutagenesis (genome editing) demonstrated in Arabidopsis
2009	Jens Boch and colleagues describe the genome targeting abilities of TAL (transcription activator–like) effectors
2010	The CRISPR/Cas 'immune system' of bacteria and archaea is described by Horvath and Barrangou

Figure 2.1. Timeline of developments in genetics and recombinant DNA technology.

new molecules. This was known as recombinant DNA technology. It meant that DNA from any source could be cloned (i.e. multiple, identical copies could be made) and bulked up in bacteria for analysis. The bacterium of choice for this purpose was and is *Escherichia coli*. This is a human gut bacterium, although the strains used in the laboratory have been modified so that they are not pathogenic. The cutting and splicing of plant, animal and fungal genes into plasmids and their propagation in *E. coli* underpinned the molecular analysis of gene structure and function. It also led to the first commercial use of recombinant DNA technology in the pharmaceutical industry, recombinant human insulin produced in *E. coli* being approved by the Food and Drug Administration of the USA in 1981.

The technology continued to develop. In 1983, Kary Mullis developed the polymerase chain reaction, which uses a DNA polymerase from a thermophilic bacterium. This polymerase survives boiling and has an optimum temperature for activity of 72°C, allowing for repeated heating and cooling of the reaction. Specific DNA fragments are amplified from extremely small amounts of template DNA to produce sufficient amounts for cloning or direct analysis. Then, in 1995, miniature microarrays were developed that allow the expression levels of tens of thousands of genes in an organism, organ or cell type to be measured in a single experiment. Microarrays consist of thousands of microscopic spots of DNA oligonucleotides (short DNA fragments), each matching a specific gene nucleotide sequence. These are probed with total RNA populations from the tissue or organ being investigated. The development of microarrays was dependent on the increasing genomic data that was coming into the public domain at that time. This culminated with the publication of the entire human genome nucleotide sequence in 1999 and, not far behind, that of the model plant, *Arabidopsis* (*Arabidopsis thaliana*), in 2000.

2.2. *Agrobacterium Tumefaciens*

Escherichia coli is the bacterium of choice for general DNA cloning and manipulation. Another bacterial species that contains a plasmid is the common soil bacterium, *A. tumefaciens*, a bacterium that infects wounded plant tissue and causes the disease known as crown gall (Figure 2.2). In 1977, Nester, Gordon and Chilton showed that genes

Figure 2.2.   A crown gall caused by the natural infection of a tree by *A. tumefaciens*,
Ashridge Forest, UK, 2013.

from a plasmid carried by *A. tumefaciens* were inserted into the DNA of
host plant cells as a natural part of the infection process. Effectively, the
bacterium could perform the genetic modification of host plant cells.
The plasmid was called the tumour-inducing or Ti plasmid. This obser-
vation led to *A. tumefaciens* becoming one of the most reliable and
widely used means of transferring foreign DNA (DNA from a different
organism) into plants.

During infection, the host plant cells release phenolic compounds.
These compounds interact with proteins encoded by virulence (*VIR*)
genes carried by the Ti plasmid and induce the bacteria to bind to plant
cell walls. Other *VIR* genes on the Ti plasmid then become active and
they cause a single strand of DNA to be nicked out of a region of the
Ti plasmid. This transfer DNA (T-DNA) is protected by specialised
proteins and is transferred to the adjacent plant cell where it integrates
into the plant DNA.

Once integrated into the plant DNA, genes present in the T-DNA become active. These genes encode proteins that perturb the normal hormone balance of the cell. The cell begins to grow and divide to form a tumour-like growth called the crown gall. The cells of the crown gall are not differentiated: in other words they do not develop into the specialised cells of a normal plant. They can be removed from the plant and cultured as long as they are supplied with light and nutrients and are protected from fungal and bacterial infection. A clump of these undifferentiated cells is called a callus.

All of the cells of a crown gall or a callus contain the T-DNA that was inserted into the original host cell. Indeed, the T-DNA will be inherited stably through callus culture and through generations of GM plants that are produced from the callus. The T-DNA contains another set of genes that induce the host cell to make and secrete unusual sugar and amino acid derivatives that are called opines, on which the *Agrobacterium* feeds. There are several types of opine, including nopaline and octopine, and different types are produced after infection with different strains of the bacterium. The genes that enable the bacterium to feed on the opines are also carried by the Ti plasmid.

A. tumefaciens has a close relative, *Agrobacterium rhizogenes*, that causes hairy root disease. *A. rhizogenes* uses a similar strategy to *A. tumefaciens,* and carries an Ri plasmid that functions in a similar way to the Ti plasmid but induces a different response in the host plant cell. Together, they are nature's genetic engineers and were using plant genetic modification millions of years before humans 'invented' it.

2.3. Use of *A. Tumefaciens* in Plant Genetic Modification

The natural mechanism of plant genetic modification by *A. tumefaciens* can be used to introduce any gene into a plant cell. This is because the only parts of the T-DNA that are required for the transfer process are short regions of 25 bp at each end, or border. Anything between these border regions will be transferred into the DNA of the host plant cell. Callus formation can be induced in the laboratory by infecting leaf pieces, stem sections, tuber discs or other 'explants' with *A tumefaciens*. In 1983, groups led by Schell and Van Montagu in Ghent, Schilperoort in Leiden, Chilton and Bevan in St. Louis and Cambridge, and Fraley, Rogers and Horsch, also in St. Louis, showed that bacterial antibiotic resistance genes could be inserted into the T-DNA carried on a Ti plasmid and transferred into plant cells. In the same year, Tim Hall and colleagues, working at the Agrigenetics Advanced Research Laboratory in Wisconsin and the University of Wisconsin, used this method to produce a sunflower plant carrying a seed protein gene from French bean. Not only was the gene present in every cell of the plant, but it was inherited stably and was active. This was the first report of what would later be called a genetically modified or GM plant.

Michael Bevan in Cambridge then developed the so-called binary vectors — plasmids that would replicate in both *E. coli*, in which they could be manipulated and bulked up, and *A. tumefaciens*. Binary vectors contain the left and right T-DNA borders but none of the genes present in 'wild-type' T-DNA. They are unable to induce transfer of the T-DNA into a plant cell on their own because they lack the virulence (*VIR*) genes that are required to do so. However, when present in *A. tumefaciens* together with another plasmid containing the *VIR* genes, the region of DNA between the T-DNA borders is transferred, carrying any genes that have been placed there.

After infection with *A. tumefaciens,* explants are transferred to sterile dishes containing solid plant growth medium and cells that have received the T-DNA divide to form calli (Figure 2.3(a)). Calli are then transferred onto medium that contains a plant hormone that induces them to form shoots (Figure 2.3(b)). Once a shoot with a stem has formed, it is transferred to medium that does not contain the shoot-inducing hormone. Hormones produced by the shoot itself then induce root formation and a complete plantlet is formed (Figure 2.3(c)). Up to this point, the process has to be done in sterile conditions to prevent bacterial or fungal infection. Once shoot and root are fully formed, the plantlet can be transferred to soil and treated like any other plant (Figure 2.3(d)).

Plants that have been altered genetically in this way are referred to as transformed, transgenic or GE, as well as GM. The term transgenic is favoured by scientists, but GM has been adopted most widely by non-specialists. All plant breeding, of course, involves the alteration (or, if you like, modification) of plant genes, whether it is through the selection of a naturally occurring mutant, the crossing of different varieties or even

Figure 2.3. *A. tumefaciens*-mediated transformation of potato. (a) Potato tuber discs that have been infected with *A. tumefaciens*, showing the formation of clumps of undifferentiated cells called calli. Cells within each callus contain genes that have been inserted into the potato genome by the bacterium. (b) Shoots induced to form from calli on potato leaf pieces by the application of a plant hormone. (c) A complete GM potato plantlet. (d) Once the plantlets are large enough, they can be transferred to soil in pots and grown in containment in a greenhouse.

related species, or the artificial induction of random mutations through chemical or radiation mutagenesis. Nevertheless, the term GM is now used specifically to describe plants produced by the artificial insertion of a single gene or small group of genes into its DNA. Genetic modification has been an extremely valuable tool in plant genetic research as well as in crop plant breeding.

2.4. Transformation of Protoplasts

Protoplasts are plant cells without a cell wall. The cells are usually derived from leaf tissue and are incubated with enzymes (cellulases, pectinases and hemicellulases) that digest away the cell wall. *A. tumefaciens* will infect and transform protoplasts; the protoplasts can then be cultured and induced to form calli by the application of plant hormones (auxins and cytokinins). GM plants can then be regenerated from the calli as described for *A. tumefaciens*-mediated transformation of explant material.

Protoplasts can also be induced to take up DNA directly. This process is called direct gene transfer or DNA-mediated gene transfer. There are two widely used methods for achieving this. The first involves treatment with polyethylene glycol (PEG) or a similar polyvalent cation. The exact way in which PEG works is not known, but it is believed to act by causing the DNA to come out of solution and by eliminating charge repulsion. The second method is called electroporation and involves subjecting the protoplasts to a high voltage pulse of electricity. This causes the formation of pores in the plant cell membrane and although these must be repaired very rapidly to prevent the protoplasts from dying, it briefly allows DNA to be taken up by the protoplasts.

Direct gene transfer into protoplasts is most commonly used as a research tool. A gene of interest is introduced into protoplasts so that its activity and function can be studied. The gene does not integrate into the protoplasts' own DNA and is eventually broken down, so the protoplasts are only temporarily, or transiently, rather than stably transformed. However, in a small proportion of the protoplasts, the introduced DNA will integrate into the host DNA and the protoplast will be stably transformed. The protoplast can then be induced to form callus and a GM plant can be regenerated from it.

There are two major drawbacks to the production of GM plants by this method. Firstly, it is not possible to induce protoplasts of all plant species to form calli and regenerate a whole plant. Secondly, the introduced DNA is often rearranged and does not function as expected.

2.5. Particle Bombardment

Flowering plants can be subdivided into two classes, monocotyledonous and dicotyledonous, depending on their embryo structure. In the wild, *A. tumefaciens* only infects dicotyledonous plants, and although progress has since been made in adapting its use for the genetic modification of monocotyledonous plants in the laboratory, and it is now the method of choice for some monocotyledonous species, its use was limited at first to dicotyledonous plants. This was important because cereals, including major crop species such as wheat, maize and rice, belong to the monocotyledonous class and these species are also not amenable to the regeneration of whole plants from protoplasts.

Transformation of cereals eventually became possible with the invention of the particle bombardment method. In this method, plant cells are bombarded with tiny particles coated with DNA. In some cases, the wall of the cell is penetrated without the cell being killed. Some of the DNA is washed off the particles and becomes integrated into the plant genome. This is carried out in a particle gun, the first of which used a small explosive charge to bombard the plant cells with tungsten particles. Subsequently, a variety of devices were developed, but one using a burst of pressurised helium gas in place of the explosive charge has been the most successful and gold particles are now used instead of tungsten. There are now hand-held versions of this device on the market, allowing DNA to be delivered to cells within intact plants. This has become an extremely useful tool for studying the transient activity and function of genes that are introduced into plant cells and remain there for a short period of time but do not integrate into the host plant DNA.

Particle bombardment, which has acquired the name of biolistics, has been particularly successful in the production of GM cereals, including maize, wheat, barley, rice, rye and oat, and the first commercial GM cereal crops were produced using this method. The plant tissues that are used are either isolated explants (Figure 2.4(a)) that are bombarded, induced to become embryogenic and regenerated, or embryogenic cell cultures. Whole plants are then regenerated from single embryonic cells that have taken up the DNA (Figures 2.4(b)–2.4(d)).

(a)   (b)

(c)   (d)

Figure 2.4. Transformation of wheat by particle bombardment. (a) Wheat embryos are the targets for bombardment with tiny gold particles coated with DNA. Some of the cells within the embryos take up the DNA and incorporate it within their own DNA. (b) The GM cells are induced to form callus material and then shoots by the addition of plant hormones. (c) Shoots are placed on a different medium lacking hormones and begin to form roots. (d) Once whole plants have been produced, they can be treated in the same way as non-GM wheat plants. This picture shows a field-trial of GM wheat at Long Ashton Research Station near Bristol, UK, in the 1990s. Pictures kindly provided by Pilar Barcelo (a)–(c) and Peter Shewry (d).

2.6. Other Direct Gene Transfer Methods

The technique of electroporation can be applied to intact cells in tissue pieces or in suspension, as well as to protoplasts, but has only been shown to work efficiently in a few species. The other direct gene transfer method to have been developed is silicon carbide fibre vortexing. Plant cells are suspended in a medium containing DNA and microscopic silicon carbide fibres. The suspension is vortexed and the fibres penetrate the plant cell walls, allowing the DNA to enter.

2.7. *Agrobacterium*-mediated Transformation Without Tissue Culture

The methods of plant transformation through the infection of explants or protoplasts with *A. tumefaciens* and the regeneration of intact plants is relatively straightforward with many plant species, including major crops such as soybean, potato and oilseed rape. It does have the drawback, however, along with the direct gene transfer methods, that the regeneration of plants from single cells often gives rise to mutation. Plants carrying harmful mutations have to be screened out before the GM plant is incorporated into a breeding programme. If the genetic modification is being made in order to investigate the function of a particular gene for research purposes, the effects of the genetic modification may be difficult to distinguish from the effects of random mutations.

For this reason, and in order to speed up the process, methods of plant transformation that do not require tissue culture have been developed. The most successful of these is floral dip transformation. This was developed by Georges Pelletier, Nicole Bechtold and Jeff Ellis using the model plant, *Arabidopsis*. Plants at the early stages of flowering are placed in a suspension of *A. tumefaciens* in a vacuum jar, a vacuum is applied to remove air surrounding the plant tissue and allow the bacteria to come into contact with plant cells, and the plants are grown to seed. Typically, approximately 1% of the seeds are GM. This method is now widely used in basic research using *Arabidopsis* and has been adapted with some success for use with other plant species, including soybean and rice.

2.8. Selectable Marker Genes

A limitation to all of the plant genetic modification techniques described above is that only some of the cells in the target tissue are GM, irrespective of the method of DNA transfer. It is, therefore, necessary to select out all of the cells or regenerating plants that are not modified, and this requires that the presence of the gene of interest can be selected for, in other words the gene must make the host plant distinguishable in some way from plants that do not carry it. Alternatively, the gene of interest must be accompanied by at least one other gene that acts as a selectable marker. Most commonly, the regeneration of GM plants is carried out in the presence of a selective agent, tolerance of which is imparted by the introduced gene or, more often, the accompanying selectable marker gene.

In practice, selectable marker genes make the transformed cells and GM plant resistant to an antibiotic (Figure 2.5(a)) or tolerant of a herbicide (Figure 2.5(b)). Commonly used antibiotic resistance marker genes confer resistance to kanamycin, geneticin or paromycin (collectively known as aminoglycosides) or to hygromycin, none of which have critical, if any use in medicine. Kanamycin, geneticin and paromycin resistance is imparted by a gene called *NPTII*, which encodes the enzyme neomycin phosphotransferase, while hygromycin resistance is imparted by one of *HPT*, *HPH* or *APH-IV*, which encode hygromycin phosphotransferases. Antibiotic resistance genes are widespread in nature, and all of these genes were obtained originally from the common gut bacterium, *E. coli*. The *NPTII* gene is probably the most widely used selective marker gene in the genetic modification of dicotyledonous plants. However, it is not generally used in the genetic modification of cereals because kanamycin is not toxic enough to cereal cells and plants.

The use of antibiotic resistance marker genes in crop biotechnology is a controversial issue (Section 5.13). However, antibiotic resistance marker genes remain an extremely useful tool in plant genetic modification for research purposes. In contrast, GM crops with herbicide tolerance are now widely used commercially (Section 3.5). The use of herbicide tolerance genes as selectable markers to accompany other genes of interest was developed because the removal of non-GM cells and regenerating plants with antibiotics was found to be unreliable with cereals. Herbicide

(a)

(b)

Figure 2.5.   Antibiotic resistance and herbicide tolerance marker genes. Plants that have been GM have to be separated from plants that have not. If the gene that has been engineered into the plant cannot be selected for, it must be accompanied by a marker gene that can. The most common marker genes impart resistance to an antibiotic or tolerance of a herbicide. (a) A GM tobacco plant (right) carrying a gene that imparts resistance to an antibiotic, kanamycin, and an unmodified plant (left) growing in the presence of the antibiotic. (b) A GM wheat plant (right) carrying a gene that imparts tolerance of a herbicide, bialaphos, and an unmodified plant (left) in the presence of the herbicide. Picture kindly provided by Pilar Barcelo.

tolerance genes work either by encoding an insensitive version of a protein that is targeted by the herbicide, or by encoding an enzyme that breaks the herbicide down.

The most widely used herbicide tolerance genes for research purposes are the phosphinothrycin acetyl transferase (*PAT* or *BAR*) genes, from the bacteria *Streptomyces hygroscopicus* and *Streptomyces viridochromogenes*. They impart tolerance to herbicides based on phosphinothrycin, including gluphosinate and bialaphos, which inhibit the enzyme glutamine synthetase. Glutamine synthetase is required for the incorporation of inorganic nitrogen into amino acids, and therefore for protein synthesis, so its inhibition results in the death of the plant. The herbicide tolerance genes encode enzymes that convert the herbicide to a non-toxic compound: phosphinothrycin acetyl transferase, for example, converts phosphinothrycin to acetylphosphinothrycin (Figure 2.6).

This selection method works for all of the cereals that have been transformed to date and as well as its use as a selectable marker system in

Figure 2.6. The action of gluphosinate on amino acid synthesis and the detoxifying action of PAT.

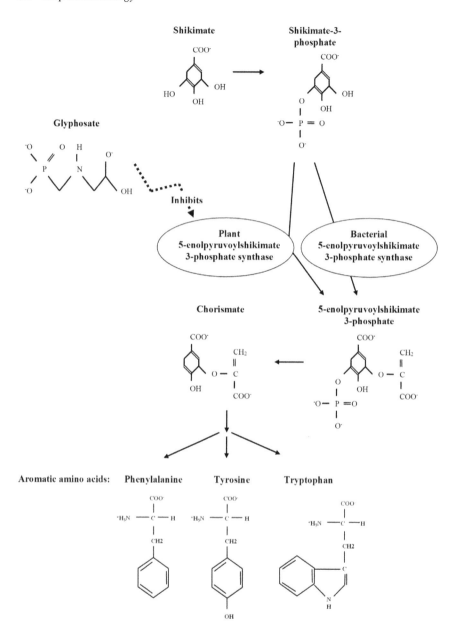

Figure 2.7. The action of glyphosate on the shikimate pathway. Glyphosate inhibits plant EPSPS, but does not affect the bacterial form of the enzyme.

genetic research, it has also been used to produce some commercial, herbicide-tolerant GM crop varieties (Section 3.5.2). Less widely used for research but also with applications in commercial GM crop varieties are genes that impart tolerance of glyphosate, such as the 5′-enolpyruvylshikimate phosphate synthase (*EPSPS)* and glyphosate oxidoreductase (*GOX*) genes from *Agrobacterium*. The enzyme EPSPS catalyses the formation of 5-enolpyruvoylshikimate 3-phosphate (EPSP) from phosphoenolpyruvate (PEP) and shikimate 3-phosphate (S3P). This reaction is the penultimate step in the shikimate pathway (Figure 2.7), which results in the formation of chorismate, which in turn is required for the synthesis of many aromatic plant metabolites including the amino acids phenylalanine, tyrosine and tryptophan. Its inhibition is therefore fatal to the plant. The shikimate pathway is not present in animals, which have to acquire phenylalanine, tyrosine and tryptophan in their diet (hence, these amino acids are referred to as essential amino acids); for this reason, glyphosate has low toxicity for animals. The *EPSPS* gene from *A. tumefaciens* encodes an EPSPS that is not affected by glyphosate, so this system bypasses the action of the herbicide, rather than breaking the herbicide down (Figure 2.7). On the other hand, GOX, an enzyme encoded by the *GOX* gene, also from *A. tumefaciens*, breaks glyphosate down to aminomethylphosphonic acid.

Finally, there are genes that impart tolerance of sulfonylurea compounds such as chlorsulfuron, including the acetolactate synthase (*ALS*) gene from maize.

2.9. Visual/Scoreable Marker Genes

A marker gene that encodes a visual or scoreable product may be used alongside a selectable marker gene in order to allow GM cells to be visualised. Examples of genes used in this way are a bacterial gene, *UidA* (commonly called Gus) (Figures 2.8(a) and 2.8(b)), a jellyfish green fluorescent protein (GFP) gene (Figure 2.8(c)), red fluorescent protein (DSred) from coral (Figure 2.8(d)) and luciferase (*Lux*) genes.

The *UidA* gene encodes an enzyme called β-glucuronidase. When supplied with 4-methyl umbelliferyl glucuronide as substrate, this enzyme will produce 4-methyl umbelliferone, a substance that fluoresces under ultraviolet light at a wavelength of 365 nm, emitting light at a wavelength of 455 nm. The intensity of fluorescence can be measured using a fluorimeter, giving an indication of the amount of product and therefore the amount of enzyme that is being produced and the activity of the gene. Plant tissue is ground up in this experiment to allow the enzyme and

(a)   (b)

(c)   (d)

Figure 2.8.

substrate to come together. An alternative substrate, X-Gluc (5-bromo-4-chloro-3-indolyl-β-D-glucuronide) can be used to penetrate thin sections of plant tissue. The β-glucuronidase enzyme produces a blue product from this substrate, and the blue colour can be seen when the tissue is examined through a microscope, allowing the exact location of gene activity to be pinpointed (Figures 2.8(a) and 2.8(b)).

The *UidA* gene has proved extremely valuable as a research tool. Its main drawback is that it is destructive; the plant tissue is killed by the assay. The products of the GFP, DS Red and luciferase genes, on the other hand, can be visualised without killing the plant tissue and have little or no toxicity for the plant cell. The GFP gene comes from an intensely luminescent jellyfish, *Aequorea victoria*. Light-emitting granules are present in clusters of cells around the margin of the jellyfish umbrella and contain two proteins, aequorin, which emits blue–green light, and GFP, which accepts energy from aequorin and re-emits it as green light. GFP fluoresces maximally when excited at 400 nm and it is visualised in GM plants using a fluorescence stereomicroscope (Figure 2.8(c)). The DsRed gene comes from coral (*Discosoma* spp.) and works in a similar way to GFP but emits red light instead of green (Figure 2.8(d)). The favoured

◄ ——————————————————————————————

Figure 2.8. (*Figure on facing page*) *UidA* **(Gus), GFP and Red Fluorescent Protein (DsRed) visible marker genes. (a) and (b) The** *UidA* **gene is bacterial in origin and encodes an enzyme called β-glucuronidase. When supplied with the appropriate substrate, this enzyme will produce a blue product, allowing the exact location of gene activity to be pinpointed. In these pictures of wheat grain sections, the gene is under the control of a wheat gene promoter that is active only in the seed endosperm (the major seed storage tissue in cereals and the white flour fraction of the grain). The pictures contrast the blue-stained endosperm cells with the unstained cells of the surrounding tissues, including the single cell-thick aleurone layer that surrounds the endosperm and is clearly visible in (b). The picture in (a) was kindly provided by Caroline Sparks, Rothamsted Research. (c) The product of the GFP gene from the jellyfish** *A. victoria* **fluoresces when excited at 400 nm and is visualised using a fluorescence stereomicroscope. The picture shows bright fluorescent foci in a wheat embryo that has been bombarded with a GFP gene controlled by a wheat promoter that is active in this tissue. Picture kindly provided by Sophie Laurie. (d) Bright red foci in a wheat embryo that has been bombarded with a DsRed gene from coral (***Discosoma*** spp.). Picture kindly provided by Caroline Sparks, Rothamsted Research.**

luciferase system actually requires two genes, *LuxA* and *LuxB*, from a bacterium, *Vibrio harveyi*. Activity of the two genes and the combining of the two gene products results in bioluminescence within the tissue, and this can be detected and measured by light-sensitive equipment.

While they are undoubtedly powerful tools in GM research, the presence of visual/scoreable marker genes in a commercial crop variety is not essential, and it is accepted that they should be avoided when producing GM plants for food or animal feed.

2.10. Design and Construction of Genes for Introduction into Plants

The coding sequence of a gene that is sourced from a bacterium, an animal or a different plant species will usually be translated properly to make a protein in a GM plant. However, the regulatory sequences within a gene that determine when and where in the organism the gene is active (Sections 1.3 and 1.4) are less likely to be recognised the more distantly related the source organism is from the GM plant. Hence, a bacterial gene introduced unchanged into a GM plant will not be active at all, and *vice versa*. This means that the coding region of the gene to be introduced into a plant is usually spliced together with regulatory sequences that will work in the plant to make what is called a chimaeric gene.

Figure 2.9 shows a schematic diagram of a chimaeric gene comprising a promoter from a wheat seed protein gene (*Glu-D1x*) attached to the coding sequence from the bacterial gene, *UidA*, and the terminator from the *Agrobacterium* nopaline synthase gene (*Nos*). As described in the previous section, the *UidA* coding sequence encodes an enzyme called β-glucuronidase that produces a blue pigment from a colourless substrate. The production of the blue product when the substrate is supplied shows that the enzyme is present and that the gene must be active. The *Nos* gene is one of those that is introduced into a plant during infection by wild-type *A. tumefaciens* and has evolved to function in plant cells. The terminator sequence ensures that RNA from the introduced gene is processed properly. The other regulatory sequences of the chimaeric gene are present in the promoter from the wheat gene.

Figure 2.9. **Schematic representation of a chimaeric gene.** Genes taken from one species and engineered into another are unlikely to work, particularly if the source and target species are not closely related. For this reason, chimaeric genes are constructed in the laboratory, comprising the coding region of the gene from the source species with control sequences (promoter and terminator) that will work in the target species. This figure shows a schematic diagram of such a chimaeric gene in which the coding region of the *UidA* gene from a bacterium is linked to a wheat gene promoter (*Glu-D1x*) that is active only in the seed endosperm and a terminator sequence from an *A. tumefaciens* nopaline synthase (*Nos*) gene that functions in plants. Sections of seeds from plants transformed with this gene are shown in Figures 2.8(a) and 2.8(b).

When wheat plants are GM with this gene, the gene is active in the same tissue and at the same developmental stage as the gene from which the promoter came. That is specifically in the major storage compartment of the seed (the endosperm, or white flour fraction) in the mid-term of seed development. This can be shown by sectioning seeds from the GM wheat plants and incubating the sections in a medium containing the substrate for β-glucuronidase, as shown in Figures 2.8(a) and 2.8(b). A blue colour develops in the endosperm tissue from mid-development seeds but not in sections taken from anywhere else in the plant.

2.11. Promoter Types

The *Glu-D1x* gene promoter is known as a tissue-specific and developmentally regulated promoter, being active in only one tissue type in the plant at a specific developmental stage. There are many different promoters of this type now available for plant biotechnology, and they are active in many different tissues and even cell types. For example, Figure 2.10 shows β-glucuronidase activity in the roots, anthers and egg sacs of three different GM wheat plants, each containing the *UidA* gene under the control of a different promoter. It is an excellent demonstration of the control over the activity of a gene that is available to a biotechnologist. Such

(a) (b)

(c)

Figure 2.10.   Different promoters allow biotechnologists to specify where in a plant a gene will be active. These pictures show the dark staining that indicates *UidA* gene activity in wheat plants that have been engineered with the gene under the control of different wheat promoters. (a) Activity in anther stalks; (b) activity in different parts of the root and (c) activity in different regions of the egg sacs. Pictures kindly provided by Pilar Barcelo.

control is not possible with other methods in plant breeding. This tight control over promoter activity may be lost if the promoter is used in a different species, however, and this has to be tested before the promoter is used in commercial applications.

There are two other types of promoter available to the plant biotechnologist. The first type is known as constitutive, meaning that these promoters are active everywhere in the plant all of the time. In GM research, constitutive promoters are used most often to drive expression of selectable marker genes. However, the fact that they are expressed everywhere in the plant has also made them the promoter of choice for use in genes that impart herbicide tolerance or insect resistance in commercial GM crop varieties.

The most widely used constitutive promoter is derived from *Cauliflower mosaic virus* (CaMV): it is called the CaMV35S promoter, and was first characterised by Nam-Hai Chua and colleagues at Rockefeller University in the 1980s. CaMV infects plants of the cabbage family. Its DNA is a circular molecule of about 8,000 bp (8 kb), and when it enters a host cell it is transcribed in its entirety to give a 35S RNA molecule or in part to give a 19S RNA molecule. The CaMV35S promoter is therefore responsible for transcription of the entire viral genome. Note that the S in the name refers to the sedimentation coefficient, which characterises the behaviour of a molecule or particle in sedimentation processes such as centrifugation; although not dependent solely on the mass of the molecule, it was one of the first methods developed for distinguishing between DNA and RNA molecules of different sizes.

The CaMV35S promoter has been used particularly successfully in the genetic modification of dicotyledonous plants. It also works in monocotyledonous plants, including cereal crop species, but has largely been replaced for cereal work with promoters that are derived from cereal genes. These include a promoter for a maize gene, *Ubi*, encoding a protein called ubiquitin, and a rice gene, *Act1*, encoding a protein called actin. There has been some controversy over the use of the CaMV35S promoter (although not in the plant science community) because of its viral origins (Section 5.18), even though CaMV only infects plants and the CaMV35S promoter represents only a small part of the viral genome and could not in itself be infective.

The third type of promoter that a biotechnologist can use is described as inducible. Promoters of this type are not active until they are induced by something such as attack by a pathogen, grazing, an environmental stress such as heat, cold or drought, or application of a chemical. There is some interest in using pathogen- or grazing-induced promoters in GM crops that are being modified to resist pathogen or insect attack. The resistance gene introduced into the plant would then only be active when it was needed. Chemically induced promoters are used in GM research so that a gene can be switched on and off in order to determine its function. Such promoters are not yet being used in commercial GM crops.

2.11.1. *The use of GM to characterise gene promoters*

A common misconception regarding plant genetic modification is that it is entirely concerned with producing new crop varieties for agriculture and is the preserve of big business. In fact, the techniques were developed in the public sector, largely in the USA and Europe, and the technique has been an extremely valuable tool in basic plant genetic research. The types of experiments to which it has been applied include analyses of gene promoter activity, functional characterisation of regulatory elements within gene promoters, the determination of gene function, studies on metabolic pathways and analyses of protein structure and function.

The analysis of gene promoter activity is one way of finding out when and where in a plant a gene is active. The promoter of the gene is spliced to a reporter gene such as *UidA*, GFP, DSRed or luciferase, and the resulting chimaeric gene, known as a reporter gene construct, is introduced into a plant by genetic modification. The level and location of promoter activity can be visualised by testing for the presence of the protein produced by the reporter gene (see Figures 2.8 and 2.10).

Further experiments can be undertaken to determine what parts of the promoter are important in its activity. This process often begins with an analysis of different promoters that have similar activity in order to find short regions of DNA (elements) that have the same sequence of base pairs. These elements are likely to play a role in controlling the activity of the promoter and are called regulatory or promoter elements (Sections 1.3 and 1.4). The function of potential regulatory elements can then be tested experimentally.

A good example of this was an analysis of the promoters of seed storage protein genes of wheat, barley, rye and maize. As their name suggests, these genes encode proteins that function as part of the storage reserve in the seed, providing nutrition for the developing seedling after germination. They have evolved into a large gene family, but are believed to have arisen from a single ancestral gene, and their activity is controlled in a co-ordinated manner. They are subject to tissue-specific and developmental regulation, being expressed exclusively in the major storage compartment of the seed, the starchy endosperm (or white flour fraction), during mid and late seed development. They are also subject to nutritional regulation, responding sensitively to the availability of nitrogen and sulphur in the grain.

Since they have a common ancestry and show similar patterns of expression, it was to be expected that these genes would have regulatory sequences in common, and this turned out to be true for many of them. One group was found to contain a conserved element, 29 bp long, positioned around 300 bp upstream of the coding sequence of the gene. This was one of the first plant regulatory elements to be characterised. It was identified by Brian Forde at Rothamsted in the UK and was first called the −300 element, subsequently the prolamin box. It has the nucleotide sequence: TGACATGTAAAGTGAATAAGATGAGTCAT.

A regulatory role for the prolamin box was established experimentally by particle bombardment of cultured barley seed endosperms with promoter/*UidA* reporter gene constructs containing different regions of the promoter. These experiments also showed that the prolamin box could be subdivided into two separate elements, one (the E box) conferring tissue-specificity, the other (the N box) reducing activity of the gene at low nitrogen levels and increasing it when nitrogen levels were adequate. All of this control is imparted by the 29 bp element.

Regulatory elements that have a positive effect on gene expression are called enhancers, while those that have a negative effect on gene expression are called silencers. The E box is an example of a tissue-specific enhancer because it causes a large increase in gene expression specifically in the seed endosperm, while the N box can act as an enhancer or a silencer, depending on nitrogen availability.

2.12. Gene Over-expression and Silencing

The use of genetic modification in the analysis of gene function is based on finding out what happens when a gene is switched off when it should be active (gene silencing) or what happens if the gene is over-expressed to make more of a protein or to make it in a tissue type or at a developmental stage where it would not normally be present. Over-expression can be achieved by introducing additional copies of a gene unchanged, or by introducing a modified version of the gene in which the coding region is spliced to a more powerful promoter.

Several methods have been developed for silencing genes in plants using genetic modification, but they are now known to involve essentially the same mechanism. Historically, the first method to be developed was the antisense technique, in which a chimaeric gene is produced using part of the gene of interest spliced in reverse orientation downstream of a promoter. The promoter may derive from the same gene, but usually it is a more powerful one. When this chimaeric gene is re-inserted into the plant, it produces RNA of the reverse and complementary sequence of that produced by the endogenous gene. This so-called antisense RNA interferes with the accumulation of sense RNA from the target gene, preventing the sense RNA from acting as a template for protein synthesis. This method was developed by Don Grierson and colleagues at the University of Nottingham in collaboration with a team led by Wolfgang Schuch at what was then ICI (the seed division at ICI was one of the sections of the company that was subsequently split off as Zeneca and after a number of re-organisations and mergers became Syngenta).

Grierson and Schuch also developed the co-suppression technique, in which one or more additional copies of all or part of a gene in the correct orientation is introduced into a plant by genetic modification. This, of course, is the technique used for over-expression of a gene but in some cases it leads to gene silencing. Gene silencing by this method requires the introduced gene to be extremely similar or identical to the native gene, so can be avoided if over-expression is the desired outcome by using a gene from a not too closely related species.

Both co-suppression and antisense gene silencing have been used to produce GM plants in which the trait is stably inherited, and both techniques have found commercial application in the extension of fruit shelf-life, most famously that of tomato (Section 3.4). There may be more than one silencing mechanism involved in both of these techniques, but the major one appears to operate post-transcriptionally, causing degradation of the RNA molecule before it can act as a template for protein synthesis. Post-transcriptional gene silencing (PTGS) turns out to be a defence mechanism against virus infection; indeed, some plant viruses have genes that suppress it. It involves the production of small, antisense RNAs. Investigations into PTGS by David Baulcombe and Andrew Hamilton, then at the Sainsbury Laboratory in Norwich, led to the development of the third method for gene silencing by genetic modification, RNA interference (RNAi), and the discovery that PTGS was the unifying mechanism for the antisense, co-suppression and RNAi techniques.

In RNAi, a plant is genetically modified to synthesise a double-stranded RNA molecule derived from the target gene. This has been done in other systems by splicing part of a gene between two opposing promoters and re-introducing it into an organism. It has been achieved in plants using gene constructs in which part of the gene is spliced sequentially in a head-to-tail formation downstream of a single promoter. This causes the production of an RNA molecule that forms a hairpin loop (hpRNA); this molecule is cleaved into short, double-stranded RNA molecules by an enzyme called Dicer that is naturally present in the cell. These short RNA molecules are called short interfering RNAs (siRNAs). The siRNAs are unwound into two single-stranded molecules, one of which (the passenger strand) is degraded, while the other (the guide strand) is incorporated into a RNA-induced silencing complex (RISC). The guide strand pairs with the complementary sequence of messenger RNA from the target gene and induces cleavage by another enzyme, Argonaute, which is present in the RISC.

The fact that PTGS is a natural response to viral infection has led to the development of virus-induced gene silencing (VIGS) to suppress gene expression transiently. This involves the production of modified plant viruses carrying nucleotide sequences corresponding to the host gene to be silenced. Infection leads to synthesis of viral double-stranded RNA, as an intermediate step in the normal viral replication process. This activates the plant's anti-viral RNA silencing pathway, resulting in degradation of the messenger RNA from the target gene. Although still a relatively new technique, VIGS is proving to be a powerful tool in plant genetic research.

Another relatively new approach is to introduce a gene into the genome of a chloroplast instead of the nuclear genome. Plants in which this has been achieved are called transplastomic plants. There are two advantages to this approach: Firstly, there is the potential for very high levels of expression of the transgene, because every cell of the green parts of the plant contains thousands of chloroplasts, each of which will carry a copy of the transgene. Secondly, the chloroplast genome is only inherited maternally, so pollen from a transplastomic plant does not contain the transgene; this may be useful if it is particularly important to prevent gene flow from the transgenic plant.

2.13. Genome Editing

Genetic modification of plants is now over three decades old, and inevitably new methods in plant genetic research and breeding have continued to be developed. The availability of whole genome nucleotide sequence data for many crop species, for example, coupled with the development of genome maps based on genetic markers, and the production of mapping populations of closely related plants derived from crosses between two varieties, is accelerating what might be called traditional plant breeding. Genetic markers may be single nucleotide differences in the genetic code of different varieties (single-nucleotide polymorphisms, or SNPs) or regions in the genome where short sequences of base pairs have been inserted or deleted (INDELs). A trait in which the two parental varieties differ may segregate in a mapping population and can be associated with specific genetic markers, enabling regions of the genome that affect the trait to be identified. These regions are called quantitative trait loci (QTL) and can be used by breeders who want to integrate a trait into their breeding lines, or investigated further to identify the specific genes responsible for the trait and genetic markers that are associated with variation in the trait. These markers can then be used in marker-assisted breeding. The identification of QTL and genetic markers is invaluable for traits that are affected by multiple genes at different sites in the genome (sometimes called complex traits).

Genomic approaches such as the identification of QTL are being used alongside or independently of genetic modification. However, there are also a number of techniques that may be regarded as advances on GM, or post-GM technologies. Several of them come under the umbrella term of genome or gene editing in that they generate targeted mutations. Some of these techniques involve a GM step, but result in a plant in which foreign DNA is no longer present but a native gene has been 'edited', and since mutagenesis is an old technique (Section 1.8.2) that does not come under GM regulations, from a scientific point of view the plant that is produced in the end is not regarded as a GM plant. These technologies use specially designed oligonucleotides (oligonucleotide-directed mutagenesis, or ODM), zinc-finger nucleases (ZFN), meganucleases, transcription activator-like effector nucleases (TALENs) or a bacterial defence system based on clustered, regularly interspaced, short palindromic repeats (CRISPR) and the Cas9 nuclease.

2.13.1. *Oligonucleotide-directed mutagenesis (ODM)*

ODM uses short, synthetic DNA molecules called oligonucleotides. These are usually single-stranded and between 20 and 100 nucleotides in length, although there are variants, including double-stranded RNA–DNA oligo-nucleotides (RDOs), and longer (>200 nucleotides) single- or double-stranded DNA molecules (small fragment homologous replacement, or SFHR). The oligonucleotide is designed to be identical in nucleotide sequence to a target gene in the plant apart from a small number (usually between 1 and 5) of 'mismatches' where the nucleotide sequence is different. The oligonucleotide can be delivered into protoplasts after treatment with PEG or electroporation (Section 2.4), or into cultured plant cells by particle bombardment (Section 2.5).

Once inside the protoplast or cell, the oligonucleotide binds to the complementary sequence of nucleotides in the target gene. The plant's own DNA repair machinery then replaces the mismatched nucleotides (Figure 2.11), or creates an insertion or deletion, depending on the nature of the mismatch between the oligonucleotide and the native gene. The exact mechanisms by which this happens are not fully understood. Nevertheless, the United States Department for Agriculture and other

Target region in plant genome

```
GCCCCGACTGGAGCGGCATGCACCAGGTCGGCGACTGCTACCTCTCCCACCAG
||||||||||||||||||||||||||||||||||||||||||||||||||||||
CGGGGCTGACCTCGCCGTACGTGGTCCAGCCGCTGACGATGGAGTGGGTGGTC
```

Oligonucleotide

TGCACCAGGACGGGGACTGCTACC

DNA pairing

```
                 GGAGCGGCATGCACCAGGTCGGCGACTGCTACCTCT
        T                                            C
       C                                            C
      A                                            C
     G                                            A
GCCCCG          TGCACCAGGACGGGGACTGCTACC          CCAG
||||||          |||||||||| ||| ||||||||||          ||||
CGGGGCTGACCTCGCCGTACGTGGTCCAGCCGCTGACGATGGAGTGGGTGGTC
```

DNA repair

```
GCCCCGACTGGAGCGGCATGCACCAGGACGGGGACTGCTACCTCTCCCACCAG
||||||||||||||||||||||||||||||||||||||||||||||||||||||
CGGGGCTGACCTCGCCGTACGTGGTCCTGCCCCTGACGATGGAGTGGGTGGTC
```

Figure 2.11.   Oligonucleotide-directed mutagenesis (ODM). A short DNA molecule (oligonucleotide) is introduced into a plant cell or protoplast. The nucleotide sequence of the oligonucleotide matches that of a target gene except for a small number of mismatches (shown in red). The oligonucleotide binds to the complementary DNA strand in the target gene (DNA pairing) and the plant cell's own DNA repair machinery then replaces the mismatched nucleotides in the target gene.

regulatory authorities have classified the technology as a mutagenesis, not a GM technique.

2.13.2. *Sequence-specific nucleases: Meganucleases, zinc finger nucleases, TALENS and CRISPR-Cas9*

An alternative method of introducing mutations at specific sites in genomes also exploits the cell's own DNA repair mechanisms of homologous recombination and non-homologous end joining. Normally, these mechanisms repair double-stranded breaks in DNA caused by ionising radiation or chemical mutagens. In homologous recombination, sections of DNA around the break are removed to leave a single-stranded overhang. This then invades an unbroken molecule in a similar way to the oligonucleotide invading a double-stranded DNA molecule in Figure 2.11. There are alternative models for what follows, involving DNA synthesis and re-annealing to produce a repaired DNA molecule. In non-homologous end joining, the DNA strands on either side of the break are directly re-joined using single-stranded overhangs that often remain on the ends of double-strand breaks. Errors occur when the overhangs on either side of the break do not match up, often resulting in the loss of nucleotides and effectively a deletion in the repaired molecule. Indeed, both repair mechanisms can result in the incorporation of errors.

In targeted mutagenesis, the tendency of the cell's DNA repair machinery to introduce mutations into the DNA sequence is exploited. Usually, transgenic plants are produced expressing a sequence-specific nuclease that will introduce a double-stranded break at a specific site in the genome, in the knowledge that in some cases the misrepair of the break will result in the creation of a mutation. The transgenic plants are then screened to find individuals in which mutations in the target gene have been created. In its simplest form, the technique is used to introduce a mutation, often a small deletion, that renders the target gene dysfunctional. Such mutations are often called knockouts. Once plants in which all the copies of the target gene have been edited are identified, they can be self-pollinated to produce progeny that all carry the genome edits but in which the transgene or transgenes will segregate. Plants can then be selected that carry the edits but not the transgene.

In some plant species, such as potato, the gene expressing the sequence-specific nuclease can be expressed transiently in protoplasts (Section 2.4). Some of the plants regenerated from the protoplasts will not contain a transgene but will carry the gene edits. In other words, there is no GM step at all in the technique.

Four classes of sequence-specific nucleases are now available: meganucleases (also called homing endonucleases), ZFNs, TALENs, and CRISPR together with the CRISPR-associated protein 9 (Cas9) nuclease (CRISPR-Cas9). In meganucleases, ZFNs and TALENs, sequence-specificity is imparted by a protein domain, while in CRISPR-Cas9 it is provided by a single guide RNA (sgRNA) that associates with the nuclease.

Meganucleases are naturally occurring restriction enzymes from microbes. Restriction enzymes cut DNA at sites with a specific sequence of nucleotides. In the case of meganucleases, the DNA recognition sequence is typically from 12 to 40 bp long, meaning that the recognition sequence will not occur very often in a genome. This distinguishes meganucleases from other restriction enzymes, which recognise short sequences of typically 6 bp. Meganucleases can be engineered to recognise different target sites, so they can be directed to specific target genes. However, the DNA recognition part of the protein overlaps with the part that does the cutting, so engineering changes in target site specificity is difficult without affecting the ability of the enzyme to cut DNA. This has been overcome by some researchers using sophisticated protein modelling techniques to predict the outcome of any changes they make, but the use of meganucleases has been superseded by methods that are simpler to use.

In ZFNs and TALENS, the DNA sequence recognition part of the protein is separate from the DNA cutting part. In fact, in both cases the enzyme is created by fusing a DNA-binding domain with the catalytic domain of the *Fok*I nuclease. *Fok*I functions as a dimer, so two DNA recognition domains are separately fused with the *Fok*I nuclease domain in order to bring the two *Fok*I proteins together at the target site (Figure 2.12).

Zinc finger domains are found in the DNA-binding region of a class of transcription factors (Section 1.3) present in almost all eukaryotic

Figure 2.12. Genome editing using ZFNs or TALENS uses the *Fok*I nuclease domain fused to a DNA binding domain. The *Fok*I nuclease functions as a dimer, so separate fusion proteins are designed with DNA binding domains that will bring the two *Fok*I proteins together at the target site.

organisms. An individual 'finger' comprises 30 amino acids folded around a zinc ion. Each finger binds a specific target site of 3 bp, which means that the overall binding specificity can be engineered by combining different zinc fingers into arrays. Further changes can be made by altering the individual amino acids in a finger, so that domains can be designed to target any DNA sequence.

TALENS uses a DNA binding domain from TALEs, a family of proteins produced by plant pathogens in the genus *Xanthomonas*. During infection, TALEs enter a plant cell and activate expression of specific genes, typically resulting in the plant becoming more susceptible to colonisation by the pathogen. The DNA binding region of TALEs comprises approximately 30 tandem repeats of 33–35 amino acids, with sequence specificity imparted by just two variable amino acids called the repeat variable di-residue. Each repeat variable di-residue recognises a specific DNA sequence, so any gene can be targeted simply by assembling an array of repeats with the appropriate repeat variable di-residues.

The CRISPR-Cas9 system is a bacterial immune system that confers resistance to foreign genetic elements such as phages (viruses that infect bacteria) and provides a form of acquired immunity. CRISPR stands for clustered, regularly interspaced, short palindromic repeats, and Cas9 is the associated nuclease used in the genome editing of plants. CRISPRs comprise repeated DNA sequences of 29 nucleotides separated by variable, 32 nucleotide spacer regions derived from segments of 'captured' viral sequences. Immunity to infection is acquired when DNA from an invading virus is recognised by a Cas protein complex, which integrates a new repeat-spacer unit corresponding to the virus DNA. The new CRISPR repeat-spacer array is then transcribed into a preCRISPR RNA (pre-crRNA), which is processed into CRISPR RNA (crRNA) and guides the nuclease to destroy the DNA of the invading virus.

In the system involving the nuclease Cas9, a transactivating crRNA (tracrRNA) molecule is also required for CRISPR activation: tracrRNA is complementary to pre-crRNA and forms a crRNA/tracrRNA hybrid, which guides Cas9 to its target. As the system has been developed for genome editing, it has been found that the tracrRNA and crRNA can be combined into a single molecule called the sgRNA. The sgRNA can be designed to target any gene sequence adjacent to a protospacer-adjacent motif (PAM) comprising three nucleotides with the sequence NGG, in which N can be any of the four bases. The Cas9 nuclease is inactive until it interacts with a sgRNA. Once it interacts with the sgRNA and is guided to its target, it anchors itself at the adjacent PAM, and cleaves the DNA.

Chapter 3

The Use of Genetically Modified (GM) and Genome-edited Crops in Agriculture

3.1. Why Use Genetic Modification (GM) or Genome Editing in Plant Breeding?

The graph shown in Figure 1.7 gives an indication of how successful plant breeding has been over the last half century. So why do plant breeders need GM or genome editing? The answer is that both techniques allow plant breeders to do some things that are not possible by other techniques. This does not mean that GM and genome editing will replace older techniques in plant breeding, far from it, but they are undoubtedly powerful tools for plant breeders to use. The advantages that GM has over other techniques are as follows:

1. It allows genes to be introduced into a crop plant from any source. Biotechnologists can select a gene from anywhere in nature and, with some modifications (Section 2.10), make a version of it that will be active in a crop plant.

2. It is relatively precise in that single genes can be transferred. In contrast, conventional plant breeding involves the mixing of tens of thousands of genes, many of unknown function, from different parent lines, while radiation and chemical mutagenesis introduce random genetic changes with unpredictable consequences.
3. Genes can be designed to be active at different stages of a plant's development or in specific organs, tissues or cell types (Figure 2.10).
4. Specific changes can be made to a gene to change the properties of the protein that it encodes.
5. The nature and safety of the protein produced by a gene can be studied before the gene is used in a GM programme (Chapter 4).

Genome editing is still in its infancy, but its obvious advantage is its precision and the ability to target a specific gene or family of genes while having no impact on the rest of the genome. Given the controversy and over-regulation that has dogged the development of GM crops, genome editing also has the advantage that it can be used to produce plants that to date have not been considered by regulatory authorities to be GM, even if there is a GM step involved in the process (Section 2.13). The caveat to that is that the European Commission has not yet issued its opinion on the regulatory status of genome-edited crops.

Both GM and genome editing require extensive background knowledge of a gene, the protein it encodes and the other genes and proteins that interact with it. This requires a significant investment of time and money compared, for example, with the generation of random mutations. Once produced, GM varieties must undergo much more detailed analysis and testing than new non-GM varieties, particularly in Europe. This probably explains why the few GM crop varieties available in Europe are essentially spin-offs from varieties developed elsewhere. The barriers to developing GM crop varieties specifically for the European market are too great. Even outside Europe, the GM crops that are currently available are based on a handful of strategies, although the number of different GM applications is growing.

Typically, development of a new GM variety takes at least 10 years and is estimated to cost $100 million (Figure 3.1). The cost escalates through the process, which means that it will always be tempting for chief

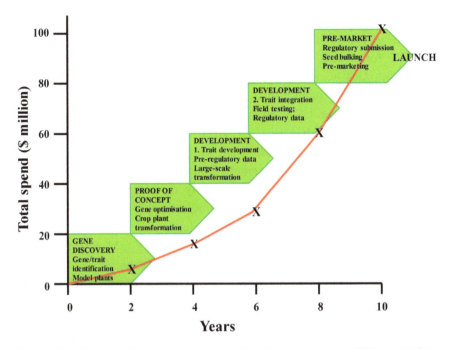

Figure 3.1. Flowchart showing the stages of development of a new GM crop variety, and graphical representation of the typical time and expense involved.

executives to call a halt, and regulatory compliance adds significantly to the cost. Establishing a market is also a problem, and some GM varieties have come and gone without doing so.

3.2. Global Adoption Rates of GM Crops

Despite these obstacles, some of the GM crops that have made it onto the market have been staggeringly successful. Data on the global use of GM crops has been compiled for several years by Clive James of the International Service for the Acquisition of Agri-Biotech Applications (ISAAA; www.isaaa.org). It is almost impossible to check the ISAAA's data for some countries, but for countries where independent data is available from other sources, the USA for example, the datasets are consistent. According to the ISAAA, the worldwide area of land planted with GM crops in 2016 was 185 million hectares (457 million acres), representing approximately 12% of total world agriculture. More than three-quarters (78%) of global soybean production was GM, with figures of 64%, 26% and 24% for cotton, maize and oilseed rape, respectively. Other crops with some GM varieties being grown commercially included sugar beet, alfalfa, papaya and poplar. GM crops were grown in 26 countries (Table 3.1), with the USA, Brazil, Argentina, Canada, India, Paraguay, Pakistan, China, South Africa, Uruguay and Bolivia all planting more than a million hectares of GM crops.

There are some notable absentees from these lists. There is currently no GM wheat being grown commercially on a significant scale, for example, and very little GM rice. China has approved some GM rice varieties for cultivation, but so far the adoption of these varieties has been either very low or zero, while any prospect of GM wheat cultivation seems even more remote. It looked like a herbicide-tolerant GM wheat variety was going to be launched by Monsanto in 2004, but the company performed an abrupt about-face and pulled out of wheat research and development altogether. No other company looks set to launch a GM wheat variety in the near future, and although Monsanto has renewed its interest in wheat it does not appear to have any GM wheat varieties ready for commercialisation.

3.3. First- and Second-generation GM Varieties

GM varieties with deliberately altered composition are sometimes called 'second generation', or described as having 'output' traits, distinguishing them from 'first generation' varieties with 'input' traits such as herbicide tolerance or insect resistance. The 'first' and 'second generation' terms are not particularly useful, partly because the first GM varieties to be launched back in 1994 were tomatoes with improved shelf life (an output trait), and partly because GM plants may have output traits combined with well-established input traits. It is true that the first traits to come on to the market in the major commodity crops were input traits, herbicide tolerance and insect resistance, and these remain by far the most successful traits that have been introduced into crop plants by GM, and the notion of first-generation input traits and second-generation output (or quality)

Table 3.1. **GM crop areas in the 26 countries growing GM crops commercially in 2016 (data from ISAAA).**

Country	GM crop area (thousand hectares)
USA	72,900
Brazil	49,100
Argentina	23,800
Canada	11,600
India	10,800
Paraguay	3,600
Pakistan	2,900
China	2,800
South Africa	2,700
Uruguay	1,300
Bolivia	1,200
Australia	900
Philippines	800
Myanmar	300
Spain	100
Sudan	100
Mexico	100
Colombia	100
Vietnam	<100
Honduras	<100
Chile	<100
Portugal	<100
Bangladesh	<100
Costa Rica	<100
Slovakia	<100
Czech Republic	<100
Total	**185,000**

traits has stuck. The following sections describe the traits that have been launched commercially and some that are in development, starting with the first trait on the market, slow-ripening fruit.

3.4. Slow-ripening Fruit

Fruit ripening is a complex molecular and physiological process that brings about the softening of cell walls, sweetening and the production of compounds that impart colour, flavour and aroma. The process is induced by the production of a plant hormone called ethylene. The problem for growers and retailers is that ripening is followed sometimes quite rapidly by deterioration and decay and the product becomes worthless. Tomatoes and other fruits are, therefore, usually picked and transported when they are unripe. In some countries they are then sprayed with ethylene to induce ripening before sale to the consumer. However, fruit that is picked before it is ripe has less flavour than fruit picked ripe from the vine.

Biotechnologists therefore saw an opportunity in delaying the ripening and softening process in fruit. If ripening could be slowed down by interfering with ethylene production or with the processes that respond to ethylene, fruit could be left on the plant until it was ripe and full of flavour but would still be in good condition when it arrived at the supermarket shelf. Various strategies based on this principle have been pursued with many different fruits, and the technology has the potential not only to improve the produce of western farmers but also to enable farmers in tropical countries to sell fruit to customers in Europe and North America. However, although slow-ripening GM varieties of tomato and papaya are grown in China, the GM tomato varieties that have been marketed in the west have been withdrawn for one reason or another, and the application of the technology in other fruits is still some years away from commercialisation.

The first commercial varieties of GM tomatoes were engineered to inhibit the process by which fruits soften as they ripen. The GM fruits change colour and acquire aroma but are less prone to bruising and deteriorate, and decay much more slowly than normal. Fruit softening is brought about by enzymes that degrade the complex carbohydrates in the cell walls within the fruit. These carbohydrates include cellulose, which comprises linear chains of glucose molecules (i.e. a type of glucan), and pectin, which is comprised of a complex group of polysaccharides based on galacturonic acid, an oxidized form of galactose (a type of sugar), and rhamnose, which is a deoxyhexose, with various sugar side-chains. The enzymes involved include cellulases, which break down cellulose, and polygalacturonase (PG) and pectin methylesterase (PME), which are ,involved in the breakdown of pectin.

Two competing groups developed tomatoes in which PG activity had been reduced in the fruit by gene suppression (Section 2.12) at approximately the same time: Calgene in the USA, and Zeneca (as it was then known) in collaboration with Don Grierson's group at the University of Nottingham in the UK. The Calgene product was the Flavr Savr tomato, and was the first GM plant product on the market. It was introduced in 1994 but was not a commercial success and was withdrawn after less than 1 year. Calgene was subsequently acquired by Monsanto, who have not so far pursued the technology further.

Zeneca chose to introduce the trait into a tomato used for processing, and this proved to be much more successful. These tomatoes (Figure 3.2) had a higher solid content than conventional varieties, reducing waste and processing costs in paste production and giving a paste of thicker consistency. This product went on the market in many countries and proved very popular in the UK where over 2 million cans of it, clearly labelled as genetically modified (Figure 3.2), were sold from its introduction in 1996 until 1999 when most retailers withdrew it in response to anti-GM hostility.

Another method for slowing down fruit ripening is to engineer a plant so that the fruit does not produce ethylene. The advantage of this strategy is that the fruit develops to the point where it would normally start to ripen and then stops, allowing the farmer to wait until all the fruit has ripened and then harvest it all at once. Ripening is then induced by spraying the fruit with ethylene. Tomatoes of this type have been developed by suppressing the gene that encodes the enzyme aminocyclopropane-1-carboxylic acid (ACC) synthase. ACC synthase is responsible for one of the steps in ethylene synthesis, and reducing levels of ACC decreases ethylene production dramatically. A tomato variety of this type was developed by a company called DNA Plant Technologies

Conventional tomatoes　　　　**GM slow-ripening tomatoes**

(a)

Tomato paste from GM tomatoes

(b)

Figure 3.2.　(a) Conventional and GM, slow-ripening tomatoes after an extended period of storage, illustrating the longer shelf-life of the GM tomatoes. (b) The most successful product from GM slow-ripening tomatoes was a tomato paste, more than 2 million cans of which were sold in the UK from 1996 to 1999. Pictures courtesy of Professor Don Grierson, University of Nottingham.

and marketed in the USA in the 1990s under the trade name 'Endless Summer'. However, the variety was withdrawn from sale because of disputes over patenting.

An alternative method with a similar outcome is to add a gene that encodes an enzyme called ACC deaminase. This enzyme interferes with ethylene production by breaking down ACC. Tomatoes of this type have been developed by Monsanto using a gene derived from a soil bacterium called *Pseudomonas chlororaphis,* but so far have not been marketed. A third method targets another of the precursors of ethylene, S-adenosyl methionine (SAM), by introducing a gene that encodes an enzyme called SAM hydrolase, which breaks SAM down; this strategy was developed by Agritope, Inc. using a viral SAM hydrolase gene, but again has not been marketed.

3.5. Herbicide Tolerance

Herbicide tolerance was described in Section 2.8 in the context of selectable marker genes, but several traits of this type have been introduced into commercial crops, and varieties carrying them have been extremely popular wherever farmers who have been allowed to use them. Weed control is an essential part of all types of agriculture, and in the developed world the method of choice for most farmers to achieve it is to spray fields with chemical herbicides. This has been true since the 1950s, long before the advent of GM. The alternatives are labour intensive, and farmers could not abandon the use of herbicides with today's labour costs and keep food prices anything close to the level that they are now. Organic farmers do not use herbicides, but organic farmers in the developed world are selling into niche markets, not providing the general population with affordable food. In the UK, for example, organic farmers simply avoid growing crops like sugar beet that are particularly sensitive to weed competition.

Most herbicides are selective in the types of plant that they kill, and a farmer has to use an herbicide that is tolerated by the crop that he/she is growing but kills the problem weeds. Commercial (non-GM) sugar beet varieties grown in the UK, for example, are tolerant of around 18 commercially available herbicides. By using a combination of different herbicides (typically eight) at different times in the season when different weeds become a problem, the farmer can protect his crop. Without weed control, sugar beet yield falls by about three-quarters and the crop is not worth harvesting.

This provides the farmer with a number of problems: the herbicide regime may be complicated; some of the herbicides may have to go into the ground before planting and weed problems that arise after the crop seed has germinated cannot be responded to; some of the herbicides involved may be toxic to humans, dangerous to handle and require protective clothing and equipment to be used; equipment and labour is required and this adds cost; finally, some herbicides persist in the soil from one season to the next, making crop rotation difficult. Many of these problems have been overcome by the introduction of GM crops that tolerate broad-range herbicides. The first of these to be introduced were Roundup-Ready soybeans, which were produced by Monsanto and have been marketed since 1996.

3.5.1. *Glyphosate tolerance*

Roundup is Monsanto's trade name for glyphosate (Figure 2.7), a broad-range herbicide that was introduced as a commercial product by Monsanto in 1974 (predating GM crops by two decades). It is now marketed under many different trade names in many agricultural and garden products. Glyphosate is non-selective and prior to its use in combination with GM plants was used primarily to clear fields completely or to remove weeds from pathways. It does not persist long in the soil because it is broken down by microorganisms; how long it persists depends on soil type and typically ranges from a few days to several months. This means that many farmers can clear their fields and plant a crop a few weeks later. It would be extremely rare for glyphosate to persist at effective levels from one season to the next.

Glyphosate is taken up through the foliage of a plant, so it is effective after weeds have become established. Its mode of action is described in Section 2.8 and Figure 2.7. It can be applied 'over the top' after the crop is established, and is transported around the plant through the phloem, so the whole of the plant dies. Its very low toxicity to insects, birds, fish and mammals, including man, means that farmers have always been very comfortable in using it.

Biotechnologists saw an opportunity to enable farmers to use glyphosate to control weeds simply and safely on growing crops by genetically modifying crop plants to tolerate the herbicide. They concentrated their efforts on EPSPS, the enzyme targeted by the herbicide (Section 2.8), initially trying to overcome the effect of glyphosate by engineering plants to over-produce the enzyme. Real success came with the discovery of a variant form of EPSPS (class II) in strains of the soil bacteria, *A. tumefaciens* and *Achromobacter*. These enzymes worked efficiently even in the presence of glyphosate. The *A. tumefaciens EPSPS* gene was isolated and inserted into soybean by particle bombardment (Section 2.5). The CaMV35S promoter (Section 2.11) was used to control the activity of the gene because the use of this constitutive promoter meant that the whole of the plant would be protected. The result was a variety of soybean that would tolerate glyphosate because, while the herbicide inhibits the plant's own EPSPS enzyme, it does not affect the enzyme produced by the introduced gene.

The first glyphosate-tolerant soybean line was field-tested in 1992 and 1993. The trait was given the trade name 'Roundup-Ready' and has since been incorporated into many soybean breeding programmes. Over 150 USA seed companies now offer varieties carrying the trait.

Glyphosate can be applied to GM glyphosate-tolerant soybeans at any time during the season. This gives farmers flexibility in weed control and the ability to respond to a weed problem if it occurs unexpectedly. The regime that is used depends on the weed pressure in a particular area. In some cases, a single application of glyphosate will control weeds throughout an entire growing season. However, the timing of the application is crucial if this is to work. Many farmers prefer to use a different herbicide before sowing in combination with a later application of glyphosate, while others use two sequential glyphosate applications.

The yield obtained with glyphosate-tolerant soybeans when they were undergoing trials was the same as for the non-GM controls. There were reports of disappointing yields in some areas when the first varieties were grown commercially, perhaps because the varieties that were available were not ideal for some locations. These reports have dried up as the trait has been bred into more varieties. There are also reports of yield increases in some areas.

The take-up of glyphosate-tolerant soybean varieties was incredibly rapid: after their introduction in 1996, their use rose to well over half of all the soybeans planted in the USA by 2001, the total area of which was approximately 30 million hectares. Adoption of GM varieties in Argentina was even more enthusiastic and reached 99% by the turn of the century. Glyphosate-tolerant soybean varieties have even been grown in Europe. Indeed, they were increasing in popularity in Romania up to 2006, although their use stopped abruptly when Romania joined the EU in 2007 because the EU had not authorized their cultivation, a situation that has not changed to this day. Nevertheless, by 2009, 77% of the global soybean crop, which covered 90 million hectares, was GM, making it difficult to source non-GM soybean. By that time, the only region where non-GM soybean could still be sourced in significant quantities was northern Brazil. However, Brazil authorised the cultivation of GM glyphosate-tolerant soybean in 2003 and is now the second largest producer of GM crops in the world, with 96.5% of its soybeans GM. Consequently, the food and feed industries in the European Union and other regions where the GM crop issue remains a controversial one can only obtain non-GM soybean by having it grown to contract, often at a premium price. Soybean meal is a major ingredient in many processed foods and an irreplaceable source of protein for animal feed. Indeed, even in Europe, the animal feed industry has been using GM soybean since the early 2000s, and tens of millions of tonnes of GM soybean are imported every year, mostly from the USA, Brazil and Argentina.

Farmers cite simpler and safer weed control and reduced costs as the main reasons for using glyphosate-tolerant crop varieties. The exact economic impact is difficult to gauge. As glyphosate-tolerant varieties became more popular, sales of glyphosate rose while those of other herbicides, such as imazethapyr, acifluorfen, bentazon and sethoxydim, fell. The price of glyphosate was then reduced to maintain competitiveness. Overall, between 1995 and 1998 there was estimated to be a reduction of $380 million in annual herbicide expenditure by soybean growers in the USA (source: National Council for Food and Agricultural Policy). However, farmers who used glyphosate-tolerant varieties had to pay a technology fee of $6 per acre, reducing the overall cost saving to $220 million. The technology fee did cause some resentment amongst USA farmers, particularly when it was not imposed on their competitors in Argentina and Brazil.

Another advantage for farmers using glyphosate-tolerant varieties is that crop rotation is made much easier. Some of the herbicides used on conventional soybean crops, such as chlorimuron, metribuzin, imazaquin and imazethapyr, remain active through to the next season and beyond. Some crops cannot be planted even 3 years after treatment. Maize, which is typically used in rotation with soybean, requires an 8–10-month gap before it can be planted. There are no such problems after glyphosate treatment because glyphosate is degraded so rapidly in the soil.

The use of glyphosate-tolerant soybean varieties has also allowed farmers to switch to a conservation tillage system, leaving the soil and weed cover undisturbed over winter. This reduces soil erosion and leaching of nitrate from the soil, where it is a fertiliser, into waterways where it is a pollutant. Conservation tillage does not require the use of herbicide-tolerant GM varieties, but the introduction of these varieties does appear to have enabled or persuaded many more farmers to switch to this system.

Last but not least, the system provides farmers with peace of mind. There are even some farmers who do not use glyphosate on their 'Roundup-Ready' soybeans at all in some seasons, preferring to use the herbicide regime that they followed before the GM variety became available. The benefit in this case is having glyphosate available 'in reserve' if a weed problem arises late in the season. This is particularly important because if weeds such as blackgrass are not controlled at this time, the soybean seed that is harvested will contain a proportion of weed seeds and buyers will pay less for it.

Glyphosate tolerance has now been engineered into many crop species, and commercial varieties of cotton, oilseed rape, maize, sugar beet and fodder beet are already on the market. It is undoubtedly the most successful GM trait to be used so far. Alternatives to the glyphosate-insensitive EPSPS enzyme are being explored. Several bacteria have been found to make an enzyme called glyphosate oxidoreductase that breaks glyphosate down and detoxifies it, and genes encoding this enzyme have been engineered into several crop species.

3.5.2. *Gluphosinate tolerance*

Glyphosate-tolerant varieties are not the only GM herbicide-tolerant varieties available. They face competition from varieties that are tolerant of another broad-range herbicide, gluphosinate (or glufosinate). As with glyphosate tolerance, this trait was discussed in Section 2.8 in the context of selectable marker genes and it is probably the most popular herbicide tolerance selectable marker trait used in plant genetic research. Crop varieties carrying the trait were developed by Plant Genome Systems, which was subsequently acquired by Aventis, which was then acquired by Bayer. Gluphosinate is marketed under the trade name Liberty. The gene used to render plants resistant to it comes from the bacterium *Streptomyces hygroscopicus* and encodes an enzyme called phosphinothrycine acetyl transferase (PAT) (Figure 2.6), which detoxifies gluphosinate. Crop varieties carrying this trait carry the trade name LibertyLink and include varieties of oilseed rape, maize, soybeans, sugar beet, fodder beet, cotton and rice. The oilseed rape variety has been particularly successful in Canada.

3.5.3. *Bromoxynil tolerance*

Another other broad-range herbicide tolerance GM trait that has been used in commercial crops confers tolerance to oxynil herbicides such as bromoxynil (3,5-dibromo-4-hydroxybenzonitrile) (Figure 3.3). Bromoxynil and other oxynil herbicides inhibit photosynthesis in dicotyledonous plants by blocking electron flow during the light reaction. This causes the production of superoxide, a highly reactive free radical, also known as a reactive oxygen species (ROS), leading to the destruction of cell membranes, inhibition of chlorophyll formation and death of the plant. Tolerance is imparted by the *bxn* gene from the bacterium *Klebsiella pneumoniae*. This gene encodes a nitrilase enzyme that detoxifies the herbicide (Figure 3.3).

A bromoxynil-tolerant oilseed rape variety was developed by Aventis and marketed in Canada in the 1990s under the name Westar Oxy-235. This variety was withdrawn in 2002 by Bayer after the acquisition of Aventis, and currently no bromoxynil-tolerant varieties of any crop are on the market. The presence of adventitious bromoxynil-tolerant oilseed rape

Figure 3.3. Bromoxynil tolerance: bromoxynil inhibits photosynthesis, and the tolerance trait is engineered into plants with the introduction of the *bxn* gene from bacterium *K. pneumoniae*. This gene encodes a nitrilase enzyme that detoxifies the herbicide. For gluphosinate and glyphosate tolerance see Figures 2.6 and 2.7.

plants in Canada has been monitored since the withdrawal of the commercial variety and is a model for the fate of GM traits in agriculture and the wider environment after they have been withdrawn. The frequency of detection has declined steadily.

3.5.4. *Combined glyphosate and dicamba tolerance*

Glyphosate-tolerant GM crop varieties have been incredibly successful, resulting in large swathes of the USA, Brazil, Argentina and other countries being planted with crop varieties that are treated with the same herbicide. This has raised concerns that herbicide-tolerant GM crops could become victims of their own success through the development of tolerance to the herbicide in weed populations. This could arise through cross-pollination between the crop and related weed species, the risk of which varies with different crop species, depending on whether or not they naturally out-cross and whether or not they have wild relatives in the area.

It could also arise as a result of the selective pressure on weed populations to develop resistance resulting from the widespread use of a single herbicide. Resistance to a particular herbicide becoming prevalent in weed populations would not lead to the 'superweeds' predicted by some anti-GM NGOs because farmers would control resistant weeds with different herbicides. However, it would lead to the herbicide-tolerant crop and its associated herbicide no longer being of benefit.

There have been some reports of glyphosate-tolerant weeds appearing in the USA, and although these are localised and sporadic, Monsanto has responded by developing soybean and cotton varieties in which glyphosate tolerance is stacked with a trait imparting tolerance to another herbicide, dicamba (3,6-dichloro-2-methoxybenzoic acid), which affects a wide range of dicotyledonous plant species. The tolerance trait is imparted by a gene from a soil bacterium, *Pseudomonas maltophilia*, which encodes an enzyme that converts dicamba to 3,6-dichlorosalicylic acid (3,6-DCSA), a chemical with no herbicidal activity. The thinking behind stacking herbicide traits in this way is that it will make it less likely that weeds will be able to develop resistance to the herbicide regime. The GM varieties are Roundup Ready 2 Xtend® soybeans and Bollgard II® XtendFlex® cotton (also containing a Bt gene (Section 3.6)), and they can be sprayed with a glyphosate/dicamba mixture.

The roll-out of the 'Roundup-Ready Xtend Crop System' that uses this technology is still ongoing in the USA, and has met some resistance from state authorities due to the volatility of dicamba and its tendency to drift onto neighbouring crops and wild areas. In response, Monsanto has developed 'Roundup Xtend™ herbicide with VaporGrip® Technology', incorporating a low-volatility dicamba, but at the time of writing, Arkansas and Missouri have imposed short-term bans on the use of dicamba while they assess the risk of damage to non-target crops and native plants, and other states may follow their lead.

3.5.5. *Herbicide tolerance through genome editing*

A herbicide tolerance trait has been developed in oilseed rape (Canola) using oligonucleotide-directed mutagenesis (ODM), a genome editing technique. The trait was developed by Cibus using its own version of

ODM, which it calls Gene Repair Oligonucleotide (GRON) technology. It is designed for use in combination with BASF's CLEARFIELD® Production System, which is based on imidazolinone herbicides. These herbicides inhibit the action of an enzyme called acetolactate synthase (ALS) (also known as acetohydroxyacid synthase (AHAS)), which is required for the synthesis of amino acids with branched side-chains (valine, isoleucine and leucine). ALS is also targeted by another class of herbicides, called sulfonylureas. These are structurally quite different to the imidazolinone herbicides, but the fact that they target the same enzyme means that they have some similarities, and the imidazolinone-tolerance trait also imparts a degree of tolerance to sulfonylurea herbicides.

The trait was produced by using ODM to introduce single-nucleotide mutations in two *ALS* genes in the plant so that the ALS enzymes encoded by those genes remained active but could no longer be bound by the inhibiting herbicide. In 2014, Health Canada notified Cibus that it had no objection to the food use of the oilseed rape (Canola) variety carrying the trait, effectively clearing the variety for commercial cultivation and making it the first genome-edited variety to be grown commercially.

3.6. Insect Resistance

Another universal problem for farmers all over the world is loss of their crops to insect grazing. An extreme example of this is the Colorado beetle, which in bad years destroys 70% of the potato harvest in Eastern Europe and Russia. Farmers in developed countries usually use chemical controls and, to a lesser extent, biological controls (such as natural predatory species) to prevent these sorts of losses. However, these measures are expensive and the pesticides used are often toxic and difficult to handle safely, while biological controls are never 100% effective.

A pesticide that is popular with organic and salad farmers is derived from a naturally occurring soil bacterium called *Bacillus thuringiensis*. This bacterium was discovered in 1901 by a Japanese scientist, called Shigetane Ishiwatari, as the cause of a disease in silkworms, but the name *thuringiensis* was registered in 1911 by a German scientist, Ernst Berliner, who noted the bacterium's action against a Mediterranean flour moth.

B. thuringiensis produces a protein that is toxic to some insects, and was first adopted as a pesticide by farmers in the 1920s, using formulations based on bacterial spores. Today, Bt pesticides are applied as powders, granules, or aqueous and oil-based liquids. Organic farmers use Bt pesticides because they are biological products that degrade rapidly, while salad farmers use them because they can be applied immediately before harvest. They have no toxicity to mammals, birds or fish and have an extremely good safety record. Disadvantages are that they are specific for insect types and do not remain effective for long after application.

The protein that *B. thuringiensis* produces that is toxic to insects is called the Cry (crystal) protein. Different strains of the bacterium produce different versions of the protein, and these versions are classified into groups, CryI–CryIV. Each group is subdivided further into subgroups A, B, C, etc. The different proteins are effective against different types of insects. CryI proteins, for example, are effective against the larvae of butterflies and moths, while CryIII proteins are effective against beetles.

Biotechnologists took the view that engineering crops to produce the Cry protein would be more efficient than applying the protein externally. This would overcome the problem of rapid loss of activity after application and would also have the advantage that only insects eating the crop would be affected. *Cry* genes from several strains of *B. thuringiensis* have now been introduced into cotton, maize, aubergine (also known as eggplant, brinjal (India and Bangladesh) or talong (Philippines)), poplar, potato, rice, soybean, and tomato, and over 23 million hectares of these crops were grown worldwide in 2016. Varieties carrying these genes are generally referred to as Bt varieties, although different companies market them under a variety of trade names.

3.6.1. *Bt cotton*

The effect of the use of Bt cotton has been particularly striking. Conventional cotton is very susceptible to insect damage, and one-quarter of USA insecticide production is used on this one crop. In areas of severe pressure from the three major pests affected by the CryIA protein (tobacco budworm, cotton bollworm and pink bollworm), Bt cotton on average requires 15–20% of the insecticide used on conventional cotton and take-up of Bt varieties in these areas has been very high. In Alabama, for example, Bt varieties accounted for 77% of the cotton planted in the first year that they became available. In areas where Cry1A-controlled pests are less abundant, adoption rates of the new cotton varieties have been low.

The use of Bt cotton pushed India up to 5th in the table of GM crop-growing countries (Table 3.1). The first Bt variety was launched in 2002 in a joint venture by an Indian seed company, Maharashtra Hybrid Seeds Co. (Mahyco), and Monsanto. Subsequently Bt 'events' have been registered by JK Agri Genetics Ltd. (JKAL) of Hyderabad, Nath Seeds of Aurangabad, the Central Institute for Cotton Research/University of Agricultural Sciences (UAS) Dharwad, and Metahelix Life Sciences of Bangalore, and by 2009 over 180 different hybrid varieties carrying the trait were available.

The adoption of Bt varieties by Indian cotton farmers increased rapidly over that period, with the cultivated area rising from 50,000 hectares in 2002 to 8.4 million hectares in 2009 (Figure 3.4), by which time Bt varieties represented just over 81% of cotton cultivation in the country.

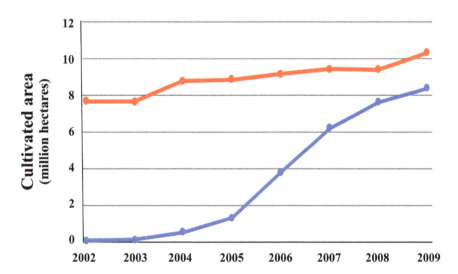

Figure 3.4.   Adoption of Bt cotton in India from 2002 to 2009 (data source: ISAAA). The total cotton cultivated area is shown in red, and the Bt cotton cultivated area in blue.

Cotton production and yield in India from 1991 to 2013 are shown graphically in Figure 3.5. These data come from the United Nations Food and Agriculture Organisation (FAO) and are therefore totally independent. They show steep rises in both yield and production coinciding with the period when farmers were adopting Bt varieties, and while other factors may have played some part, the inescapable conclusion from these data and the enthusiastic adoption of Bt varieties by farmers is that Bt varieties outperformed their non-GM competitors by a considerable margin.

3.6.2. *Bt maize*

The effects of adoption of Bt maize in the USA have varied from area to area. The principle maize pest affected by the Cry1A protein is the corn borer, and where the corn borer is not particularly abundant, USA farmers do not spray against it, preferring to tolerate any losses incurred if they suffer an infestation. In these areas, use of Bt maize led to an increase in yield typically of 10–15%. In areas where the corn borer is abundant, farmers have had to spray heavily and pre-emptively with insecticide to control it. In these areas, use of Bt maize did not increase yields but did lead to a decrease in the number of sprays applied per season from an average of 7–8 to an average of 1–2. In areas where the corn borer is not present, the Bt varieties are not used because the seed is more expensive than conventional seed.

The use of Bt crops in the USA is regulated by the Environmental Protection Agency (EPA), and the EPA has been concerned right from the start about the possibility of insects developing resistance to the Bt protein. This would make not only GM Bt crops useless but also the Bt pesticide used by organic and salad farmers. The EPA therefore insisted that farmers must provide refuges of non-Bt crops where insects that developed resistance to Bt would be at a selective disadvantage. This has not been popular with farmers but appears to have been successful in preventing resistance from developing.

An unexpected benefit of using Bt maize varieties is that the Bt grain contains lower amounts of fungal toxins (mycotoxins). These compounds, which include potent carcinogens such as aflatoxin and fumicosin, are a particular problem in tropical countries because of the warm and humid

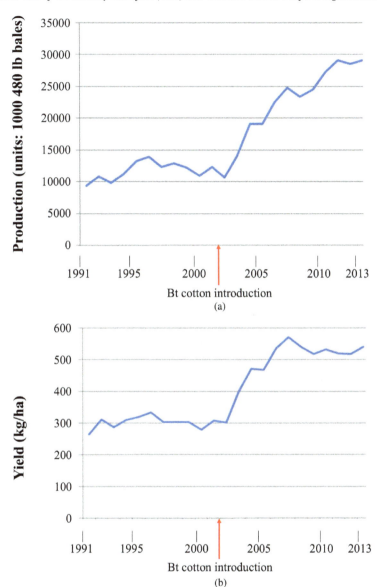

Figure 3.5. **Graphs of UN Food and Agriculture Organisation (FAOSTAT) data on Indian cotton production (a) and yield (b) from 1991 to 2013, showing the date of the introduction of Bt cotton.**

conditions that maize grain is stored in, and are associated with a high incidence of throat cancer. In temperate countries they are considered to pose most risk to grain-fed animals and the people most aware of them are horse-owners. Nevertheless, there is undoubtedly a low-level presence of these chemicals in the human food chain, and the European Commission has set maximum levels for their presence in foodstuffs since 2006. The reduced insect damage in Bt crops means that they are less susceptible to fungal infection and therefore are less likely to contain mycotoxins. This has been an important selling point in South Africa, and adoption rates for Bt maize in that country are over 80%.

Bt maize is currently the only GM crop to be grown commercially in Europe. Indeed, maize variety MON810, which carries the Bt trait, is one of only two GM crop varieties to be licensed for cultivation in the European Union, the other being the Amflora potato variety (Section 3.9). European legislation on GM crop planting was tightened in 1999, but Bt maize had already become established in Spain before then and continues to this day, the area of Bt maize cultivation in Spain in 2016 being 129,081 hectares. This accounted for most of the GM crop area in Europe in 2016, which was 136,363 hectares, increasing from 57,287 hectares in 2006 (Figure 3.6). These figures, of course, are tiny compared with those for global GM crop cultivation.

The other European country growing more than a thousand hectares of Bt maize in 2016 was Portugal, with just over 7,000 hectares, while Slovakia and the Czech Republic both grew tiny areas. The variety has also been grown in France, Poland, Germany and Romania in the past. Indeed, it appeared to be gaining ground amid great controversy in France in 2006 and 2007; one French farmer was quoted as saying that Bt seed cost him about $48 an acre, instead of about $38 for regular seed, but that he was saving between $24 and $48 an acre because he was not spraying pesticides to kill corn borers. The French government banned cultivation of Bt corn in 2008, despite the fact that the variety had been approved for cultivation throughout the EU (Section 4.4), putting it at loggerheads with the European Commission.

There is little data available on the impact of Bt maize in Spain, but the more than two decades of its cultivation have seen maize yields in

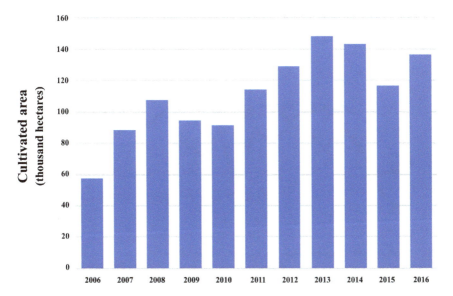

Figure 3.6. **GM crop planting (entirely Bt maize variety MON810) in the European Union from 2006 to 2016. Data: Eurostat.**

Spain overtake those of its major European competitors, France and Italy (Figure 3.7).

3.6.3. *Bt potato*

The other *Cry* gene that has been used in plant biotechnology is the *CryIIIA* gene of *B. thuringiensis* var. *tenebrionsis*. The CryIIIA protein is effective against beetles, and potato varieties containing the *CryIIIA* gene are resistant to infestation by the Colorado beetle, a pest that can devastate potato crops. One such variety, NewLeaf, produced by Monsanto, was on the market in the USA for several years in the 1990s but was withdrawn due to poor sales. USA potato plantations are attacked by a number of pests that are not controlled by Bt, in addition to the Colorado beetle, and farmers have turned to new, broad-range insecticides instead of the GM option. The NewLeaf variety also failed to find favour with fast-food chains, a key market for potatoes in North America. Nevertheless,

Figure 3.7.   Maize yield in the three most important maize-growing countries of the European Union, France (red), Italy (blue) and Spain (yellow), from 1994 to 2015. Of the three, only Spain permits the cultivation of Bt maize variety MON810, despite the European Commission granting permission for the variety to be cultivated through-out the European Union. Data: Eurostat.

Bt potato may have a role to play in Eastern Europe and Russia where the Colorado beetle is a huge problem.

3.7. Virus Resistance

Plant viruses could be regarded as being as dangerous to humans as viruses that cause the worst human diseases. Viruses such as *Cassava mosaic virus* and the *Feathery mottle virus* of sweet potato, for example, are responsible for the deaths of millions of people every year through the destruction of vital food crops. Farmers attempt to prevent viral plant diseases by controlling the insect pests that carry the disease. Those in developed countries also have access to virucidal chemical applications but, as with other agrochemicals, these are expensive and, once estab-lished, viral diseases are difficult to bring under control.

The first methods used by biotechnologists to engineer plants to be resistant to viruses arose from investigations into the phenomenon of cross-protection, in which infection by a mild strain of a virus induces resistance to subsequent infection by a more virulent strain. Cross protection has been known about for some time and appears to involve the coat protein of the virus because it is lost if a virus is encapsulated in the coat protein of a different virus. Engineering a plant to make a viral coat protein was found to mimic cross-protection, with the plants showing delayed or no symptoms after infection with the virus.

An example of the commercialisation of virus-resistant GM varieties incorporating this technology is a papaya grown in the Puna district of Hawaii. In 1998, after an epidemic of *Papaya ringspot virus* (PRSV) almost destroyed the industry, growers switched to a virus-resistant GM variety carrying a gene encoding a PRSV coat protein. There is no other known solution to an epidemic of PRSV, and some commentators claim that the GM variety saved the papaya industry in Hawaii. However, the decision to use the GM option remains controversial even to this day because some markets were lost as a result.

Another method to engineer virus resistance into plants is to use gene suppression techniques (Section 2.12) to block the activity of viral genes when the virus infects. Monsanto, for example, targeted a replicase gene from *Potato leaf role virus* (PLRV) in this way to induce resistance to PLRV in potato. A potato variety containing this trait and the Bt insect-resistance trait was marketed under the trade name NewLeaf Plus but, like the NewLeaf variety, this was not commercially successful and was withdrawn.

More successfully, scientists from the Brazilian Agricultural Research Corporation (EMBRAPA), the research arm of the Ministry of Agriculture, have produced a GM pinto bean (*Phaseolus vulgaris*) (called common bean in the UK) that is resistant to the *Golden mosaic virus*, which can devastate bean plantations. The GM bean uses RNA interference (RNAi) to prevent replication of the virus. The variety was approved on 15th September 2011 by the Brazilian National Technical Commission on Biosafety (CTNBio), and released for cultivation in 2014. Common bean is an extremely important crop for subsistence farmers in Brazil. Similar technology is being applied to induce resistance to *Spotted wilt virus* in tomato.

Virus-resistant papaya, tomato and sweet pepper have been approved for commercial cultivation in China. China has a huge population to feed and over half the population is dependent on agriculture for its livelihood, so the use of plant biotechnology there is extremely important. However, perhaps even more exciting is the potential of this technology in developing countries where losses to viral diseases are the greatest and have the most severe consequences. Kenya, for example, has field-tested a GM sweet potato variety engineered to be resistant to *Feathery mottle virus*. This virus is estimated to reduce Kenya's sweet potato production by half, and the yield of the GM variety was reported to be 80% higher than non-GM varieties in the trials. However, these trials were conducted over a decade ago and no varieties have yet been made available to farmers. GM cassava that is resistant to *Cassava mosaic virus* may be available sooner. Several teams have addressed this problem, and the Bill and Melinda Gates Foundation and other donors have provided substantial funding to an international consortium of laboratories, enabling real progress to be made. East African cassava plantations are being threatened by another disease, *Brown streak virus*, so the situation is urgent. However, adoption of GM crops in Africa has been tortuously slow, and virus-resistant crops have still not been approved for commercial cultivation anywhere on the continent.

3.8. Modified Oil Content

The principle components of plant oils are triacylglycerols, which are made up of three fatty acid molecules attached to glycerol (Figure 3.8). Fatty acids comprise a carboxylic acid (COOH) group at the end of a hydrocarbon chain, which is generally regarded as including at least eight carbon atoms (the 18-carbon stearic acid is shown in Figure 3.8). Naturally occurring fatty acids are synthesised from the two-carbon acetyl group of acetyl-CoA (CH_3–CO–S–CoA) (Figure 3.8), and therefore have an even number of carbon atoms.

In seeds, the carbon that is required for fatty acid synthesis comes from sucrose imported from photosynthesising leaves (Figure 3.9). The sucrose is hydrolysed by invertase to form glucose and fructose, which are then phosphorylated to produce hexose phosphates. The hexose phosphates then enter glycolysis and are converted through multiple steps to the three-carbon molecule, pyruvate. The pyruvate is converted to acetyl-CoA in plastids by a multi-protein complex called the pyruvate dehydrogenase complex. Acetyl-CoA has many potential uses in the cell, and the first committed step in fatty acid synthesis is its carboxylation to form

Figure 3.8. Diagrams showing the structure of acetyl-CoA, the 18-carbon saturated fatty acid, stearic acid and triacylglycerol (R1, R2 and R3 represent alkyl chains (chains of carbon and hydrogen atoms) derived from fatty acids). From: Halford, N. G. (2015) *An Introduction to Bioenergy*. Imperial College Press, London. ISBN: 978-1-78326-623-4 (hardcover). ISBN: 978-1-78326-624-1 (softcover).

malonyl-CoA (COO^-–CH–CO–S–CoA). This reaction is catalysed by the enzyme acetyl-CoA carboxylase and requires one molecule of bicarbonate (CHO_3^-) and one molecule of ATP. Malonyl-CoA is used in fatty acid biosynthesis by an enzyme complex called fatty acid synthase. The process is multi-step, involving transfer of the acyl group (RCO^-, where R is a hydocarbon chain), the sequential addition of two-carbon units and finally termination of the reaction. Another protein, acyl carrier protein (ACP), is involved in the intermediate reactions. The predominant products are the 16- and 18-carbon saturated fatty acids, palmitic and stearic acid (Figure 3.10). These are exported from the plastid and interact with coenzyme A (CoA) and acyl-CoA binding protein (Figure 3.9). The resulting complex is transported to the endoplasmic reticulum where the

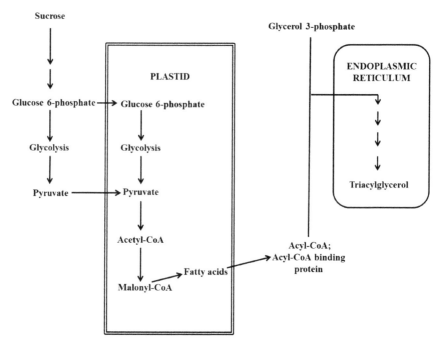

Figure 3.9. Diagram showing the compartmentation of fatty acid synthesis and assembly into triacylglycerols. From: Halford, N. G. (2015) *An Introduction to Bioenergy*. Imperial College Press, London. ISBN: 978-1-78326-623-4 (hardcover). ISBN: 978-1-78326-624-1 (softcover).

fatty acids are assembled into triacylglycerols. During or after this process, the fatty acid chains may be lengthened or processed to produce the range of fatty acids present in plant oils.

Different fatty acids are distinguished not only by chain length but also by the number and position of double bonds between the carbon atoms in the chain. Carbon atoms joined by double bonds (C=C) are described as unsaturated, while those that are joined by single bonds (C–C) are saturated. Fatty acids in which all of the carbons are joined by single bonds are similarly called saturated fats, while those with one double bond are monounsaturates and those with more than one are polyunsaturates. Plant oils contain a variety of fatty acids with different chain lengths and degrees of saturation (Figure 3.10 and Table 3.2), and the various properties of oils from different plants are determined by their differing fatty acid contents.

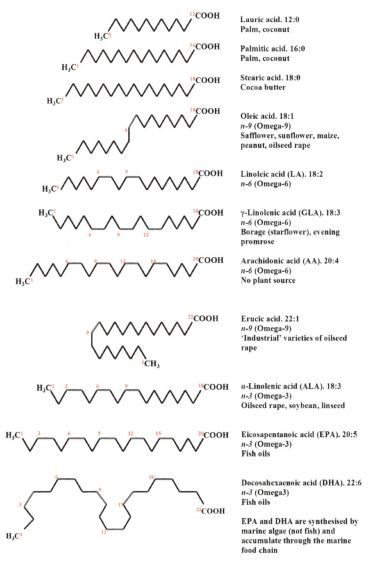

Figure 3.10. Structures of some fatty acids and their sources. Carbon chains are represented by the zigzag lines, with a carbon atom at each point in the chain. Carbon atoms linked by single bonds to adjacent atoms have two hydrogen atoms attached, while those participating in double bonds have one. Numbers in red indicate the position of a carbon atom in the chain, taking the carbon at the methyl (omega) end as 1, and indicate the first and last carbon, as well as the first desaturated carbon participating in each double bond.

Table 3.2. Typical fatty acid content (%) of some plant oils. Where the total does not equal 100%, the balance is made up of minor constituents that are not shown. OR = oilseed rape.

	Caprylic acid 8:0	Capric acid 10:0	Lauric acid 12:0	Myristic acid 14:0	Palmitic acid 16:0	Stearic acid 18:0	Oleic acid 18:1 Omega-9	Linoleic acid; LA 18:2 Omega-6	α-Linolenic acid; ALA 18:3 Omega-3	γ-Linolenic acid; GLA 18:3 Omega-6	Erucic acid 22:1 Omega-9
Oilseed rape (Edible)					4	2	61	21	11		
High lauric GM OR			40		3	1	40	15			
Oilseed rape (Industrial)				1	3	1	14	11	11		54
Sunflower					9	7	10	74			
Palm				1	44	5	39				
Soybean					11	4	23	54	7		
Vistive GM soybean					11	4	80				
Coconut	8	7	48	18	9	3	6				
Linseed					6	4	20	16	54		
Evening primrose					6	2	8	70		10	
Starflower					10	4	8	37		23	
Maize					12	2	30	54	1		
Peanut						18	47	29			
Olive					14	2.5	69	12			

The number of carbon atoms and double bonds in the fatty acid chain is given as a ratio; oleic acid, for example, is an 18-carbon fatty acid with one double bond (a monounsaturate) (Figure 3.10), so the ratio is 18:1. The position of the first double bond in the chain is given in the form *n–x*, where *x* is the position of the first unsaturated carbon with respect to the methyl (omega) end of the molecule (Figure 3.10); so, for oleic acid, this number is *n*–9. Oleic acid (18:1, *n*–9) is also known as an omega-9 fatty acid for this reason. Oleic acid is the major constituent of olive and oilseed rape oil.

Fatty acids have a high energy content and the molecules pack together to form high-density energy stores in the fatty tissues of animals and the seed oils of plants. Their high energy content makes them suitable for conversion to biodiesel and use as a transport fuel. Triacylglycerols (Figure 3.8) are esters of fatty acids and glycerol, esters being compounds formed by the condensation of a fatty acid with an alcohol. The most common method of biodiesel manufacture is through transmethylation (also known as Fischer esterification or transesterification with methanol), which produces individual fatty acid methyl esters (FAMEs) with glycerol as a by-product. Ethanol may be used in the reaction instead of methanol to produce fatty acid ethyl esters. Fatty acids can also be reacted with sodium hydroxide to produce soap. Stearic acid, its methyl and ethyl esters, and, for comparison, sodium stearate (the soap derived from stearic acid) are shown in Figure 3.11.

Well-known plant fatty acids (Figure 3.10 and Table 3.2) include lauric acid (12:0) and palmitic acid (16:0) which are the prevalent fatty acids in coconut and palm kernel oil. Lauric acid is used in the manufacture of cosmetics and detergents: for example, the sodium lauryl sulphate that is commonly used in shampoos is derived from it. Palmitic acid is used in both detergents and foods. Stearic acid (18:0) is a major component of cocoa butter, while oleic acid (18:1, *n*–9) is the major constituent of olive and oilseed rape oil. Linoleic acid (LA) is found in safflower, sunflower and maize oil, and makes up about 20% of oilseed rape oil. Gamma linolenic acid (GLA) (18:3, *n*–6) is another omega-6 fatty acid and is identical to LA except that it has an additional double bond at *n*–12. Starflower (borage) oil contains more GLA than any other plant oil, but evening primrose oil is also a good source and in the UK, at least, is generally much cheaper.

Stearic acid

Stearic acid methyl ester

Stearic acid ethyl ester

Sodium stearate (soap)

Figure 3.11. Diagrams representing the structures of stearic acid, a common 18-carbon, saturated fatty acid, the fatty acid methyl ester (FAME) and ethyl-ester that are derived from it during the manufacture of biodiesel, and sodium stearate, the soap derived from stearic acid. From: Halford, N. G. (2015) *An Introduction to Bioenergy.* Imperial College Press, London. ISBN: 978-1-78326-623-4 (hardcover). ISBN: 978-1-78326-624-1 (softcover).

Animal fat is made up predominantly of stearic acid and oleic acid. The terms 'oil' and 'fat' are to some extent interchangeable, although oil is more commonly used for liquids while fats is used for solids. The relatively high saturated fatty acid content of animal fats means that they are generally solid at room temperature, while the high proportion of polyunsaturated fatty acids (PUFAs) in plant oils means that they are generally liquid.

Nutritionally, fatty acids are important because they are constituents of cell membranes. They can also be converted into hormones and other compounds with vital functions in the human body. As a result, deficiencies in some fatty acids are linked with a range of medical conditions and diseases. Fatty acids that the human body requires but cannot synthesise itself are described as essential: i.e. they are an essential part of the human diet. LA is an example of an essential fatty acid, and GLA is also often described as essential because, although the human body can make GLA, it requires LA to do it; it cannot make GLA *de novo*. Lack of these fatty acids causes dry hair, hair loss and poor wound healing. One of their uses is to be converted to arachidonic acid (AA; 20:4, *n*–6) (Figure 3.10), with GLA being an intermediate between LA and AA (this is significant because most people consume more LA than GLA). AA is present in the phospholipids of cell membranes and is abundant in the brain and muscles.

Long-chain omega-6 PUFA (LC-PUFA)s like these are also used to make a family of compounds called eicosanoids, which include prostaglandins, leukotrienes and isoprostanes. These compounds perform a number of essential physiological functions including regulation of the immune system, blood clotting, neurotransmission and regulation of cholesterol metabolism. Dietary intake of GLA may be particularly important for people with diabetes, who appear to convert LA to GLA inefficiently. AA itself is found in meat, eggs and dairy products, but there is none whatsoever in a vegan diet. It is therefore particularly important for vegans that they consume adequate amounts of LA or GLA from plant sources. Many animals, including the domestic cat and other felines, are unable to synthesise AA from LA or GLA at all and consequently are obligate carnivores.

Erucic acid (22:0) is another interesting plant fatty acid. It is poisonous but has a variety of industrial uses. It has similar properties to mineral oils, but is readily biodegradable, and is used in transmission oils, oil paints, photographic film and paper emulsions, healthcare products, plastics (in the form of its derivative, erucamide) and biodiesel. Erucic acid used to make up about 50% of oilseed rape oil, making the oil unfit for human consumption. Oilseed rape also had the problem that the meal left over after oil extraction contained high levels of compounds called glucosinolates, which are also poisonous and have a bitter taste. Oilseed rape was used as a forage crop in the UK in the 19th Century, but the first time that it was grown widely in the UK was during the Second World War, when it was used to produce industrial oil. Plant breeding of different varieties over the next 30 years reduced the levels of erucic acid and glucosinolates to the point where these varieties were considered acceptable for human consumption. This process involved an intense programme of mutagenesis (Section 1.8.2).

The first low erucic acid, low glucosinolate varieties were grown in Canada in 1968. Nevertheless, oilseed rape did not get its seal of approval for human consumption (Generally Recognized as Safe) from the US Food and Drug Administration until 1985. Canadian producers then came up with the name Canola (Canadian oil low erucic acid) for edible oilseed rape oil. This name was adopted all over North America not only for the edible oil but also for the crop itself.

The oil of modern oilseed rape varieties for human and animal consumption is made up of oleic acid (60%), LA (20%) and the omega-3 PUFA, alpha linolenic acid (ALA) (18:3, *n*–3) (10%), with palmitic, stearic and other fatty acids together accounting for the other 10%. High erucic acid varieties are still grown today for industrial purposes but are not permitted to be placed in the food chain. The rapidly increasing demand for biodiesel may lead to more cultivation of high erucic acid varieties and conflicting demands on plant breeders and growers for varieties with different oil profiles suitable for different end uses.

3.8.1. *High lauric acid oilseed rape*

Oilseed rape oil has already been a target for biotechnologists, partly because the oil is one of the cheapest edible oils on the market, so growers and processors are always likely to be interested in anything that adds value to it. The company Calgene, subsequently taken over by Monsanto, genetically modified an oilseed rape variety to produce high levels of lauric acid in its oil. This variety was introduced onto the market in 1995. It contains a gene from the Californian Bay plant that encodes an enzyme that causes premature chain-termination of growing fatty acid chains. The result is an accumulation of lauric acid, a 12-carbon, saturated fatty acid, to approximately 40% of the total oil content, compared with 0.1% in unmodified oilseed rape, making the oil an alternative to palm and coconut oil as a source of lauric acid for detergents. However, palm oil production in particular is very efficient, and oil from the GM oilseed rape did not gain a foothold in the market. Cultivation of this variety was never more than small-scale and has now ceased.

3.8.2. *High oleic acid soybean*

The other major crop that has been genetically modified to increase the value of its oil is soybean. In conventional soybean, oleic acid (18:1) is synthesised in the plastid and transported to the endoplasmic reticulum as oleoyl-ACP (Figure 3.9). It is converted to linoleic acid (18:2) in the endoplasmic reticulum by a single desaturation step carried out by an enzyme called a Δ12-desaturase, encoded by the gene *FAD2* (fatty acid desaturase-2) (note that, somewhat confusingly, desaturases are usually named according to the position of the first carbon in the double bond that they introduce with respect to the acidic or delta end of the molecule, rather than the omega end). Palmitic acid (16:0) is also released from the plastid, and this process is controlled by an enzyme called palmitoyl-ACP thioesterase. More palmitoyl-ACP thioesterase activity results in more palmitoyl-ACP being released from the plastid and less being converted to stearoyl-ACP and oleoyl-ACP. Palmitoyl-ACP thioesterase is encoded by a gene called *FATB*.

A GM soybean line was developed in the early 1990s with reduced *FAD2* gene expression and consequently less Δ12-desaturase activity. It accumulates oleic acid to approximately 85% of its total oil content, compared with approximately 20% in non-GM varieties. It also contains less palmitic acid. This variety was produced by PBI, a subsidiary of DuPont, and is marketed under the trade name 'Vistive'. A second modification that targets the *FATB* gene has since been developed, and oil from soybean with reduced expression of both *FATB* and *FAD2* is 95% oleic acid. Monsanto have also produced a high oleic acid variety, marketed as 'Plenish', but while it carries a GM trait imparting tolerance of glyphosate, the high oleic acid trait results from mutagenesis, not GM.

The main reason for developing these varieties is that oleic acid is very stable during frying and cooking and less prone to oxidation during storage than polyunsaturated fats, making the oil less likely to form compounds that affect flavour. The traditional method of preventing polyunsaturated fat oxidation involves hydrogenation and this runs the risk of creating *trans* fatty acids. *Trans* fatty acids contain double bonds in a different orientation to the *cis* fatty acids present in natural plant oils. They behave like saturated fat in raising blood cholesterol and potentially contributing to blockage of arteries. Indeed, although technically unsaturated they cannot be described as such in USA food labelling. The oil produced by high oleic acid GM soybean requires less hydrogenation and consequently there is less risk of *trans* fatty acid formation. Hydrogenation also generates oils that are solid at ambient temperatures and make a biodiesel with poor low-temperature performance, high viscosity, and low lubricity. High oleic acid soybean oil should therefore be a better feedstock for biodiesel production.

Reducing *trans* fatty acid content of foods is a worthy target, but conventional soybean is also a valuable source of LA and ALA. If the GM soybean oil is used in food, consumers must be made aware of the changes in fatty acid content, with possible disadvantages made clear. However, if the GM soybean oil is used for industrial purposes such as biodiesel, this becomes a moot point. Linseed oil is an alternative source of ALA. However, it has a strong flavour and 'industrial' odour, so it is not consumed in large quantities, although it is taken by some people as a nutritional supplement.

3.8.3. *Long-chain polyunsaturated fatty acids (LC-PUFAs)*

The examples above are undoubtedly only the first of many GM crops modified to change their oil content. Many oils have uses in the plastics and industrial oil industries, and there are examples of crops that have been modified to make oils of this type. Another exciting prospect is the production of oils with nutritional or pharmaceutical (sometimes called nutraceutical) properties. As discussed above, animals, including humans, have only a limited capacity to synthesise some fatty acids and a number of fatty acids, including LA and ALA, that are produced by plants, have been identified as essential dietary components. In the case of GLA, which can only be made relatively inefficiently from LA, the plant species for which it is a major oil constituent, namely evening primrose and borage (starflower), make poor crops. An early target for biotechnologists, therefore, was to take a gene from borage that encodes a Δ-5 desaturase and engineer it into oilseed rape or sunflower, both of which accumulate LA, to convert some of the LA into GLA. This has been achieved in the model plant, *Arabidopsis*, but the technology has not been commercialised because, despite the problems in sourcing GLA from evening primrose and starflower, the cost of the oil itself makes up a relatively small part of the total cost of the product by the time it reaches the consumer. There is, therefore, no economic incentive to produce the oil more cheaply. However, further engineering of *Arabidopsis* by a team led by Johnathan Napier at Rothamsted Research and the University of Bristol in the UK has shown that it is possible to produce GM plants that accumulate the omega-6 LC-PUFA, arachadonic acid (AA).

3.8.3.1 *Engineering components of fish oils into crop plants*

Another application being pursued by Johnathan Napier's team and by several biotech companies is the production of marine fish oils in plants. Marine fish oil contains omega-3 LC-PUFAs such as eicosapentaenoic acid (EPA) (20:5) and docosahexaenoic acid (DHA) (22:6) (Figure 3.10). There is no plant source of these omega-3 LC-PUFAs. The human body can synthesise them from ALA but the efficiency is low: approximately 5% in men and slightly higher in women.

As with the omega-6 LC-PUFAs such as AA, the omega-3 LC-PUFAs are precursors for eicosanoids, but the eicosanoids derived from omega-3 and omega-6 LC-PUFAs have distinct properties. In general, eicosanoids derived from omega-6 LC-PUFAs are pro-inflammatory, pro-aggregatory and immuno-active, while those derived from omega-3 LC-PUFAs have little or no inflammatory activity and act to modulate platelet aggregation and immuno-reactivity.

Foetal development is now known to require omega-3 LC-PUFAs, and both EPA and DHA are commonly added to infant formula milk. There is also increasing evidence of the effectiveness of omega-3 LC-PUFAs in the prevention of cardiovascular disease. This link was first proposed in the 1970s as a result of studies on Inuit populations who consumed fat-rich diets but suffered almost no cardiovascular disease. The fish oils consumed by these people were rich in omega-3 LC-PUFAs, and follow-up studies supported the hypothesis that it was these compounds that were protecting the Inuit from the disease.

More recently, it has also emerged that omega-3 LC-PUFAs can protect against metabolic syndrome, a collection of pathologies that are indicative of progression towards cardiovascular disease, obesity and type 2 diabetes. Treatment for metabolic syndrome now includes dietary intervention with reduced carbohydrate intake and increased fish oil consumption.

The problem with marine fish as a source of omega-3 LC-PUFAs such as EPA and DHA is that marine fisheries are in a global crisis with stocks diminishing through over-fishing. There are also concerns about the levels of environmental pollutants such as dioxins, PCBs and heavy metals in fish oils. For example, the use of fish oils in products for babies and young children is not permitted in the USA. Furthermore, fish are no more capable of synthesising these compounds *de novo* than other animals are. In fact, omega-3 LC-PUFAs are synthesised by marine algae and accumulate through the marine food chain; in other words, fish acquire them in their diet. This means that farmed fish are not a sustainable source of these essential dietary components; they only accumulate them because they are fed fishmeal from marine fish. Finding a sustainable alternative source is therefore essential, and so GM plants producing EPA and DHA are being developed.

Engineering plants to produce these very LC-PUFAs from ALA has not been a simple task because of the numbers of genes involved (Figure 3.12). However, the team at Rothamsted Research has now produced GM lines of *Camelina sativa*, or false flax, that accumulate high levels of both EPA and DHA in their seed oil. *C. sativa* is a member of the *Brassicaceae* family, and has some history of cultivation as an oilseed crop in Europe. Oil derived from the GM *Camelina* plants has been shown to be suitable for feeding Atlantic salmon in fish farms, resulting in the

Figure 3.12.   Pathways for biosynthesis of very long chain omega-3 PUFAs that are present in fish oils, including EPA and DHA. These omega-3 LC-PUFAs are actually synthesised by marine algae but accumulate through the marine food chain. The longest omega-3 PUFA in crop plants is ALA.

accumulation of acceptable levels of EPA and DHA in the fish without the need to supply them with fish meal. The effect in mice of consuming feed enriched with oil from the GM *Camelina* has also been studied, the results showing that the benefits are similar to those derived from fish oils. Monsanto, BASF and DuPont are also working in this area, and are believed to be developing varieties of soybean and oilseed rape that produce EPA and DHA.

3.9. Modified Starch for Industrial Uses: The Amflora Potato

Starch from cereals, potato, sweet potato and cassava is the most important carbohydrate in the diets of humans and farm animals and by far the most important dietary energy source. Over 2 billion tonnes of starch are produced annually in cereal grain and 700 million tonnes in roots and tubers. In developed countries, starch provides 35% of daily caloric intake, while in developing countries it may provide 80%. Starch is also used as a thickener in the food industry, while non-food uses include the manufacturing of paper (starch typically makes up 8% of a sheet of paper), adhesives, gypsum wall boards and textile yarns.

Starch is actually composed of two components, amylose and amylopectin (Figure 3.13). Amylose comprises mainly linear chains of glucose residues linked by bonds between the first carbon of one glucose unit and

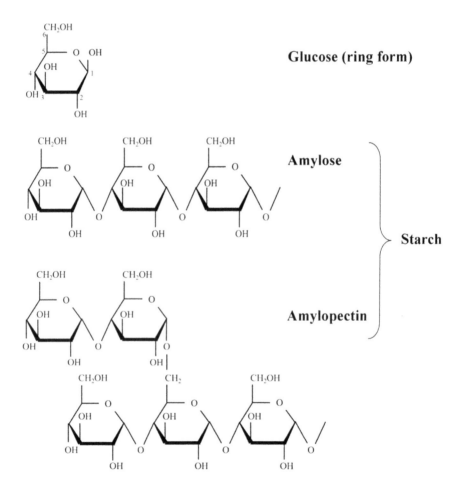

Figure 3.13.   Diagrams representing the structures of glucose in its ring form, and starch, which is made up of unbranched and branched chains of glucose molecules (amylose and amylopectin, respectively).

the fourth carbon of the next (α(1–4) linkages). Each chain may be 1,000 units long. Amylopectin, on the other hand, comprises highly branched chains in which short chains of 20 or so units are joined in a branch structure, with bonds between the first carbon of one unit and the sixth of another at each branch (α(1–6) linkages). In storage organs such as grains, storage roots and tubers, starch is made in specialised plastids (compartments) within the cell, called amyloplasts. The enzymes responsible are called starch synthases, with soluble forms of the enzyme responsible for amylopectin synthesis and enzymes within the starch granule itself (granule-bound starch synthases, or GBSS) for amylose synthesis.

A difficulty in using starch for industrial purposes or as a food additive is that amylose and amylopectin have different characteristics and must be separated or modified chemically before use. Amylose, for example, has gelling properties that are undesirable in some processes. The relative amounts of amylose and amylopectin are determined by the activities of different starch synthases and BASF developed a GM potato, marketed as 'Amflora', in which the activity of GBSS was greatly reduced by RNAi (Section 2.12). The starch from this potato variety was composed almost entirely of amylopectin, whereas normal potato starch contains approximately 20% amylose. Amflora was developed for the European market and was mired in the European Union's tortuous regulations covering the use of genetically modified crops for over a decade. It was finally approved for cultivation in 2010, making it only the second GM crop variety to be approved for cultivation in Europe (Section 4.4). BASF cultivated the Amflora variety in Germany and Sweden in 2011 in order to produce seed potatoes for sale to farmers. However, in 2012 the company announced that it was ceasing all its GM crop development in Europe and relocating it to the USA. The Amflora variety was not considered suitable for the USA market and is now not available anywhere.

3.9.1. *Modified starch through genome editing*

The potato *GBSS* gene has also been targeted by genome editing, and an international team led by Per Hofvander at the Department of Plant Breeding, Swedish University of Agricultural Sciences has reported the production of edited potato lines with full knockout of *GBSS* gene function using CRISPR-Cas9 technology (Section 2.13.2). This was one of the first reports of CRISPR-Cas9 being used in potato, and since cultivated potato is a tetraploid, full knockout means genes in all four genomes were successfully edited.

Another interesting aspect of the study was that the editing was achieved through transient transfection of isolated potato protoplasts with a plasmid carrying the genome-editing gene constructs, rather than through the generation of GM plants. Non-GM edited plants were regenerated directly from the protoplasts. This meant that there was no need to eliminate any transgenes by back-crossing.

The genome-edited lines have not been developed for commercial release, but the study showed the potential of genome-editing techniques to produce non-GM lines with desirable traits in crop species like potato in which it would be extremely difficult or impossible to achieve the same improvement by traditional breeding methods.

3.10. GM Crops for Biofuel

Starch can be used to produce sugars through enzymatic digestion. Until recently, these sugars were used largely in the food industry, but over the last few years there has been a huge increase in the use of sugars derived from starch for the production of ethanol for fuel. This industry is now well-established in the USA, where the annual growth rate in ethanol production, almost entirely from maize starch, was 25% between 2003 and 2007 (Figure 3.14). In 2010, bioethanol production took a third of the USA maize crop, a staggering amount.

In the UK, bioethanol production will use wheat grain as a feedstock. The first large bioethanol plant in the UK was built by Ensus at Wilton in the Teeside region of north-east England. Ensus is owned by two USA private equity funds, the Carlyle Group and Riverstone. It represents an investment of £300 million and went into production in February 2010. It is designed to make over 400 million litres of bioethanol and 450 million tonnes of animal feed (a valuable co-product in bioethanol production) per year from approximately 1.2 million tonnes of wheat grain, making it one of the largest biorefineries in Europe. Animal feed is not the

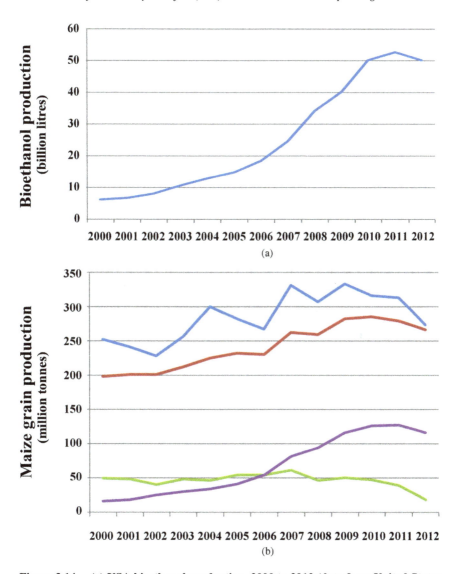

Figure 3.14. (a) USA bioethanol production, 2000 to 2012 (data from United States Energy Information Administration). (b) USA maize grain harvest (blue), consumption (red), exports (green) and used for bioethanol production (purple), 2000–2012 (data from United States Department for Agriculture). From: Halford, N. G. (2015) *An Introduction to Bioenergy.* Imperial College Press, London. ISBN: 978-1-78326-623-4 (hardcover). ISBN: 978-1-78326-624-1 (softcover).

only co-product; the 300,000 tonnes of CO_2 produced by the plant is collected for use in the manufacture of carbonated drinks.

A plant of similar size and cost (£350 million) has been built at Saltend in Hull, Humberside, also on the east coast of England, by Vivergo, owned jointly by Associated British Foods, DuPont and BP. This factory went into production in 2013 and can process 1.1 million tonnes of wheat per year and produce 420 million litres of ethanol and 500,000 tonnes of animal feed. A third major bioethanol facility at Grimsby started production in 2014. This operation is owned by Vireol and financed in part by another venture capital group, Future Capital Partners. It cost approximately £200 million and is about half the size of the other two, capable of producing annually 200 million litres of bioethanol, 175,000 tonnes of animal feed and 127,000 tonnes of CO_2 for the drinks industry from 500,000 tonnes of wheat.

Of the 15 million tonnes or so of wheat grain that have typically been produced in recent years in the UK, approximately 3 million tonnes have been exported, most of it destined for animal feed. It is this wheat that is likely to be used for bioethanol production instead. However, all three bioethanol operations have had difficulties, citing poor quality feedstock, delays in the adoption of legislation compelling fuel companies to incorporate renewables into their products, and competition from imported bioethanol as causes. If and when all three are fully operational, the Ensus, Vivergo and Vireol facilities will use 2.8 million tonnes of UK wheat, representing almost 20% of the total UK wheat harvest in an average year.

In principle, bioethanol production from cereal grains is no different to the production of alcoholic drinks by malting, fermenting and distilling. During traditional malting, cereal grains are exposed to moist, warm conditions that induce partial germination and allow the grains' own starch-degrading enzymes to be synthesised. This has been considered too complex for bioethanol production for fuel, and a 'dry-grind' process has been favoured in which the entire kernel is ground into a coarse flour, then slurried with water. The resulting mash is then cooked, treated with enzymes, fermented and distilled. The first enzyme to be added is α-amylase, an enzyme that cuts the chains of glucose units at random locations to yield shorter glucan chains. Ultimately its products are maltotriose (a trisaccharide composed of three glucose units), maltose (a disaccharide composed of two glucose units) and limit dextrin (a mixture of branched and unbranched glucans), and the process is known as gelatinisation and liquefaction. Additional enzymes, called glucoamylases, are then added to produce smaller sugars that can be fermented; this process is known as saccharification.

A key factor in the profitability of bioethanol production is how much ethanol is produced per unit of starch, sometimes referred to as the fermentability of the starch. In an effort to improve the efficiency of the process, Syngenta have produced a GM maize variety with a highly thermostable α-amylase gene, *amy797E*, from the thermophilic bacterium, *Thermococcales*. This variety is claimed to give a better yield of ethanol in the dry-grind process. It was deregulated by the USA authorities in early 2011. Producing as much ethanol as possible per unit of starch is also important because the demand for starch and starch-containing crops has been growing more rapidly than starch production for many years, culminating in stocks reaching alarmingly low levels in 2005–2006 and a consequent steep rise in prices. The situation was exacerbated by a long-term drought in Australia, one of the major wheat exporters. Farmers responded to the rise in prices by increasing production, and prices fell back again. However, 2010 saw another sharp increase in prices, this time as a result of a severe drought in Russia, another major wheat exporter, which culminated in the Russian government banning wheat exports in order to ensure that they had sufficient grain to meet domestic demand. Crop commodity prices remain volatile, and the long-term trend in demand continues upwards as more grain is used to produce bioethanol, world population expands and people in emerging countries want to eat more, better food and more meat (which means more grain for animal feed) and have the money to pay for it. Meeting that demand in the coming years will be a challenge.

3.10.1. *High lysine maize*

The use of grain starch for bioethanol production leaves a high-protein co-product that is an ideal raw material for animal feed. An example of a variety that has been produced with both co-products in mind is Mavera maize, which was developed by Renessen, a joint venture between Cargill and Monsanto. Mavera combines a triple stack of input traits: resistance to corn rootworm and the European corn borer, and tolerance of glyphosate. The novelty, however, is that the grain has a relatively high lysine content. Lysine is an essential amino acid (in other words, it must be obtained in the diet) for non-ruminant animals, including pigs, horses and poultry, and cereal grains generally do not contain enough of it. This means that animal feed must contain other sources of lysine, for example soybean or oilseed rape meal. Indeed, this is one of the reasons why home-grown barley has lost most of the UK animal feed market to imported soybean.

The high lysine trait in Mavera maize is imparted by a bacterial gene encoding a lysine-insensitive dihydrodipicolinate synthase. Dihydrodipicolinate synthase is an enzyme in the pathway for lysine synthesis that is feedback-inhibited by lysine, thereby providing the major regulatory control for flux through the pathway. The bacterial enzyme is not affected by lysine, allowing the amino acid to accumulate to levels beyond what it would in an unmodified plant.

The benefits of Mavera maize are reduced costs because of its input traits and a more nutritious feed that does not require supplementation in order to provide animals with adequate amounts of lysine. Currently, it is being grown entirely for USA domestic bioethanol production with high-lysine animal feed as a valuable co-product. It is an indication of where maize biotechnology in the USA is heading, with triple input traits and a value-adding quality trait as the cherry on top. Nothing like it is likely to be available to European farmers in the foreseeable future, and this should sound alarm bells for European maize growers who may find themselves competing with varieties like Mavera but unable to grow them themselves because of the stifling effect of EU regulations. This is discussed further in Chapters 4 and 5.

3.11. Vitamin Content: Golden Rice and Other Crops with Elevated Levels of β-Carotene

Vitamins are compounds that are required for the normal functioning of our bodies but which we cannot make ourselves. They are, therefore, an essential component of our diet. They are involved in physiological processes such as the regulation of metabolism, the production of energy from fats and carbohydrates, the formation of bones and tissues, and many others.

While nutritionists still argue over the optimum levels of different vitamins that people should consume, it is agreed that consumers in the developed world have access to foods (fresh fruit, vegetables, bread, meat, dairy products) containing all of the vitamins that they require. They also have access to an array of vitamin supplements. However, while most people eat plenty of meat and dairy products, a significant number avoid them altogether, and few people eat the recommended amounts of fresh fruit and vegetables. The health of some people might, therefore, be improved if the levels of vitamins in foods were increased. In other words, if vitamin levels were increased in foods that consumers like, instead of consumers being persuaded to eat foods that are good for them. Furthermore, for foods that consumers associate with a healthy lifestyle, such as breakfast cereals, there is market advantage to be gained in improving the nutritional value of the product and advertising the fact to consumers.

Obviously, manufacturers have the option of adding vitamins to their products, but in some cases this is expensive. This might present opportunities for biotechnologists to engineer crop plants to contain more vitamins. Folic acid, for example, deficiency of which may cause gastrointestinal disorders, anaemia and birth defects, is one possible target. Others are the fat-soluble Vitamins E and K, deficiencies in which are associated with arterial disease and, in the case of Vitamin K, post-menopausal osteoporosis.

The more obvious need for increasing the vitamin content of peoples' diets is in the developing world, where acute vitamin deficiency is the cause of poor development, disease and death. Vitamin A deficiency, for example, is common in children in developing countries who rely on rice as a staple food. It causes symptoms ranging from night blindness to those of xerophthalmia and keratomalacia. These related conditions lead to pathologic dryness of the conjunctiva and cornea. The conjunctiva may become thick and wrinkled and the cornea ulcerated, leading to total blindness. Vitamin A deficiency is the leading cause of blindness in children in the developing world: over 5 million children develop xerophthalmia annually and approximately 250,000 of these become blind. Vitamin A deficiency also exacerbates diarrhoea, respiratory diseases and measles; consequently, improving vitamin A status in children reduces death rates by 30–50%. Efforts to eradicate vitamin A deficiency are co-ordinated by the World Health Organisation but have failed so far due to the difficulty of reaching those in need.

Ingo Potrykus, a biotechnologist at the Swiss Federal Institute of Technology in Zurich, saw the possibility of using a complementary approach through the genetic modification of rice, the staple crop for many subsistence farmers. Rice grain does contain vitamin A but only in the husk, and the husk is generally discarded because it rapidly goes rancid during storage, especially in tropical countries. Potrykus aimed to produce a rice variety that would make pro-vitamin A (β-carotene, a precursor that humans can process into vitamin A) in its seed endosperm, the largest part of the seed that is eaten. A conventional breeding approach would not work because no conventional rice variety makes any pro-vitamin A at all in its endosperm.

While rice endosperm does not contain vitamin A or β-carotene, it does synthesise a compound called geranylgeranyl pyrophosphate (or geranylgeranyl diphosphate), which is an early intermediate in the pathway for β-carotene production. Potrykus' team successfully engineered the rest of the pathway (Figure 3.15) into rice, using phytoene synthase (*psy*) and lycopene β-cyclase genes from daffodil (*Narcissus pseudonarcissus*), and a phytoene desaturase (*crtI*) gene from the bacterium *Erwinia uredovora*.

The GM rice producing pro-vitamin A was crossed with another line containing high levels of available iron. Rice normally contains a molecule called phytate that ties up 95% of the iron, preventing its absorption in the gut. The GM rice contains a gene encoding an enzyme called phytase that breaks phytate down. The high pro-vitamin A/high available iron hybrid was called Golden Rice.

The aim of the Potrykus team was to enable people to obtain sufficient pro-vitamin A to avoid the symptoms of vitamin A deficiency in a daily ration of 300 grams of rice. Anti-GM NGOs claimed that much larger quantities of the rice would have to be eaten to achieve the Recommended Daily Allowance (RDA) level. In their press release of February 2001, for

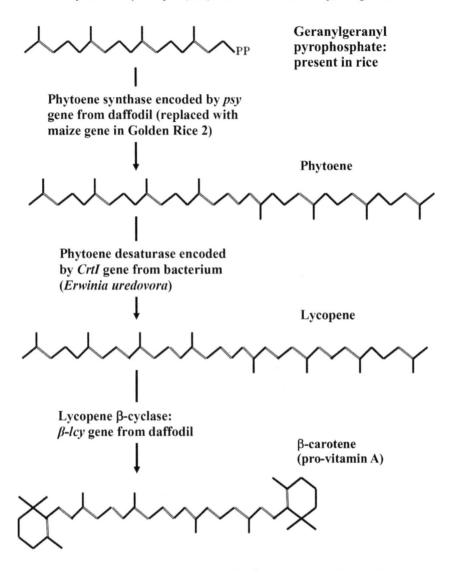

Figure 3.15. Engineering of the pathway for β-carotene (pro-vitamin A) synthesis from geranylgeranyl pyrophosphate in Golden Rice.

example, Greenpeace described Golden Rice as 'fool's gold', claiming that an adult would have to eat at least 3.7 kg of dry rice (12 times the normal intake of 300 g) to get the daily recommended amount of pro-vitamin A. The press release coincided with the start of a breeding programme undertaken by the International Rice Research Institute in the Philippines to cross Golden Rice with local rice varieties. Potrykus replied that the calculations of Greenpeace were based on RDA values for western consumers, who have access to as much vitamin A as they care to eat, and that the levels of pro-vitamin A achieved with Golden Rice were in the 20–40% range of the RDA in a 300 g serving. Nutritional experts involved in the project believed that this level would have a significant effect in preventing blindness and other symptoms associated with severe vitamin A deficiency. Even so, the full RDA would be provided by 0.75 to 1.5 kg dry weight, not 3.7 kg. The debate fizzled out anyway when Golden Rice 2 was announced. Golden Rice 2 was developed by Syngenta, in collaboration with Potrykus, by replacing the daffodil phytoene synthase gene with one from maize, and it contains many times more pro-vitamin A than Golden Rice 1.

The production of Golden Rice was first reported in 2000. At that time, it was an experimental breeding line and it had to be crossed into local varieties around the world to produce varieties that could be made available to farmers. However, governments are very protective of their export markets in Europe and Japan, where resistance to GM crops has been strong, and have also been alarmed by the anti-GM stance taken by some other countries. This has undoubtedly held up the release of varieties to farmers and, frustratingly, that has still not occurred. The process has taken many years longer than it needed to because of what Potrykus has called 'European attitudes' to GM crops. This may come back to haunt European regulators and NGOs if high pro-vitamin A rice is seen to save lives. Meanwhile, similar technology is being used in cassava and banana, both of which are staples in Africa.

3.12. Fungal and Oomycete Resistance

Fungal and oomycete diseases of plants cause severe losses in crop production. An extreme example is the Irish potato famine of the 19th Century, which was caused by a pandemic of the oomycete, *Phytophthora infestans*, the cause of late blight disease. Resistance to infection on the part of the plant is imparted in many cases by so-called resistance or R genes. R genes encode proteins that act as receptors for pathogen molecules and induce a hypersensitive response (HR) in which there is rapid cell death around the entry point of the fungus, preventing development of the disease.

The pathogen proteins that are detected by R genes are encoded by avirulence genes (*Avr* genes). An *Avr* gene aids infection of plants that do not carry an R gene that recognises it, but must be discarded by the pathogen to overcome resistance imparted by an R gene. Thus, there is a complicated relationship between the genetics of plants and their fungal pathogens that has arisen through millions of years of co-evolution.

The success of R genes is dependent on there not being too many individual plants carrying any one particular R gene. There is then some selective advantage in a pathogen retaining an *Avr* gene, in that it is better able to infect those plants without the R gene. In a field of crop plants that are essentially all the same, this selective advantage does not exist. As a result, the efforts of plant breeders to breed resistance into crop varieties by introducing particular R genes have rapidly been overcome by the emergence of new strains of the pathogen that are not affected. Nevertheless, biotechnologists believe that if they can learn more about how the R gene system works they will be able to engineer fungal and oomycete resistance into crop plants, perhaps by stacking multiple R genes in one crop variety.

Biotechnologists have also attempted to engineer resistance to late blight into potato by finding wild species of potato that are resistant to the disease, identifying the genes that impart resistance and transferring them to cultivated potato varieties by GM. In theory, moving genes from wild potato species into breeding programmes is possible with conventional breeding methods. However, this is problematic because the wild varieties are diploid, while cultivated varieties are tetraploid (in other words there has been a doubling of the chromosome number in the development of cultivated potato). This makes crossing wild and cultivated potato species difficult. Furthermore, potatoes contain toxic compounds known as glycoalkaloids, including solanine and chacocine, which deter insects and other herbivores and have antifungal properties. These large organic compounds (the molecular formula for solanine is $C_{45}H_{73}NO_{15}$) cause nausea, dizziness, vomiting, diarrhoea, heart arrhythmia and, in extreme cases, coma and death. Their levels in the tubers of cultivated potatoes have been reduced by breeders over many decades, and in the UK must comply with a regulatory limit of 200 mg per kg (although levels increase if the tubers are exposed to light and start to turn green). The tubers of wild potato species usually contain much higher levels of these compounds and are not considered fit for human consumption. If a cross is made between a wild and cultivated species, it requires a lengthy process of back-crossing to reduce the glycoalkaloid concentration back to acceptable levels. Moving genes from wild to cultivated potato species is therefore an example of where GM is clearly the most efficient and safest route.

BASF have used a gene called *RB* from a wild potato species, *Solanum bulbocastanum,* to confer broad-spectrum resistance to late blight in a cultivated potato variety. The company initially chose Ireland to field-trial the GM potato lines in 2006, but this met with fierce controversy and the plans were abandoned. In 2007, two trials were undertaken in the UK, but the trial sites were vandalised and it is not clear how much data the company were able to obtain.

In 2010, two GM potato lines were field-trialled at the John Innes Centre in Norwich. One contained the *Rpi-vnt1.1* gene from *Solanum venturii*, the other the *Rpi-mcq1* gene from *Solanum mochiquense*. Both GM lines had shown resistance to late blight in glasshouse experiments. In the field, resistance in the line carrying the *Rpi-mcq1* gene broke down, but the line carrying the *Rpi-vnt1.1* gene performed well. The *Rpi-vnt1.1* gene has now been used to confer late blight resistance in a commercial potato variety, Innate® Generation 2, produced by the J. R. Simplot Company of Boise, Idaho, USA. Innate potatoes are described in more detail in the next section on crops with reduced potential for acrylamide formation during cooking and processing.

An alternative approach to the challenge of engineering fungal resistance into crop plants is to modify them with genes that express fungicidal proteins. Examples are genes encoding the enzymes chitinase and ß-glucanase, both of which attack the cell walls of fungal hyphae as they enter the plant. Transgenic plants containing genes for these enzymes have been reported to have increased resistance to pathogenic fungi under experimental conditions.

3.13. Food Safety: Reducing the Potential for Acrylamide Formation During Cooking and Processing

Acrylamide (C_3H_5NO) is a processing contaminant produced at the high temperatures generated by frying, baking, roasting or high-temperature processing. A processing contaminant is a substance that is produced in a food when it is cooked or processed, is not present or is present at much lower concentrations in the raw, unprocessed food, and is undesirable either because it has an adverse effect on product quality or because it is potentially harmful. Acrylamide is classified as a Group 2A carcinogen (which means that it is considered to be probably cancer-causing in humans), and a report from EFSA's Expert Panel on Contaminants in the Food Chain (CONTAM) in 2015 stated that the margin of exposure (MOE) for dietary acrylamide indicated a concern for neoplastic (tumour-inducing) effects. The MOE is defined by EFSA as the ratio of the level at which a small but measurable effect is observed to the estimated exposure dose; it is also sometimes defined as the ratio of the maximum no-adverse-effect-level to the estimated exposure dose. Potato, coffee and cereal products are the major contributors to dietary acrylamide intake.

The European Commission issued 'Indicative Values' for the presence of acrylamide in food in 2011, and revised them downwards for many product types in 2013. Indicative Values are not regulatory limits or safety thresholds, but if a product is found to exceed the Indicative Value, the relevant food safety authority should take action to ensure that the manufacturer addresses the problem. Currently, the Indicative Values for potato crisps and French fries are 1,000 and 600 parts per billion (ppb), respectively, while the Indicative Value for soft bread is 80 ppb, breakfast cereal 400 ppb, biscuits 500 ppb, roast coffee 450 ppb and instant coffee 900 ppb. However, the European Commission has just approved strengthened risk management measures including compulsory Codes of Practice and the renaming of Indicative Values as Benchmark Levels, with reduced Benchmark Levels of 750 ppb for potato crisps, 500 ppb for French fries, 50 ppb for soft bread, 150 to 300 ppb for breakfast cereal, depending on its composition, 350 ppb for biscuits, 400 ppb for roast coffee and 850 ppb for instant coffee. It is going to be challenging for the food industry to meet these levels. The proposals also include a specific reference to the setting of maximum levels with regulatory enforcement for acrylamide in certain foods, stating that this should be considered following the entry into force of the new regulations. The proposals could be in force by early 2018. The European Commission's direction of travel on acrylamide regulation is, therefore, clear, and the supply and processing chains for the foods that are affected and the crops they are made from must be prepared for medium- and long-term changes in the regulatory framework on acrylamide.

Acrylamide forms predominantly *via* the deamination and decarboxylation of free (non-protein) asparagine in the Maillard reaction; a complex series of non-enzymatic reactions between free amino acids and reducing sugars, such as glucose, fructose and maltose. A reducing sugar is one that is capable of acting as a reducing agent because it has a reactive double bond in a free aldehyde group (e.g. glucose) or a free ketone group (e.g. fructose). Free asparagine and reducing sugars can therefore be regarded as acrylamide's precursors, although the carbon skeleton of acrylamide is derived entirely from asparagine.

In cereal products, the determining factor for acrylamide formation is the concentration of free asparagine. The relationship between precursor concentration and acrylamide formation in potato is more complex, with reducing sugars being the major determinants of acrylamide-forming potential in most datasets but free asparagine contributing to the variance in some. In particular, free asparagine concentration has more influence in French fry rather than crisping varieties, probably because French fry varieties contain higher concentrations of sugars. Despite this complexity, the bottom line is that acrylamide will not form if free asparagine and reducing sugars are not present.

Melanoidin (brown) pigments and complex mixtures of compounds that impart the flavours and aromas that are associated with fried, baked and roasted foods, also form within the Maillard reaction. This makes acrylamide reduction more problematic because any measures that are taken are likely to affect the characteristics that define products and brands, and are demanded by consumers, as well as the concentration of acrylamide. It suggests that free asparagine concentration would be a better target for genetic intervention than the concentration of reducing sugars, even in potato, because lowering free asparagine concentration would specifically affect acrylamide formation, while a decrease in reducing sugar concentration would affect all the products of the Maillard reaction.

The food industry has devised many strategies for reducing acrylamide formation in its products, including using appropriate varieties for specific end uses, modifying processes and improving quality control. These have been compiled in a 'Toolbox' produced by FoodDrinkEurope. The potato crisp sector can demonstrate evidence of the success of the 'Toolbox' approach in that manufacturers' data on acrylamide levels in potato crisps in Europe from 2002 to 2011 showed a reduction of 53% from 763 ppb in 2002 to 358 ppb in 2011, although acrylamide levels plateaued after that. The data also show the effect of seasonality arising from the influence of potato storage on acrylamide levels, with acrylamide in the first 6 months of the year being higher than in the second 6 months (Figure 3.16). Storage of potatoes is a critical factor, because if they are stored too warm they sprout, but if they are stored too cold they undergo cold sweetening, in which starch and sucrose are broken down, while glucose and fructose accumulate. Sweetening also occurs as the potatoes break dormancy: this is referred to as senescent sweetening.

Figure 3.16. Mean acrylamide levels in samples of potato crisps in Europe from 2002 to 2016, plotted monthly with standard errors, illustrating the seasonality in acrylamide levels. From: Powers, S. J., Mottram, D. S., Curtis, A. and Halford, N. G. (2017) Acrylamide levels in potato crisps in Europe from 2002 to 2016, *Food Additives and Contaminants: Part A*, 34, 2085–2100.

Preventing cold and senescent sweetening would enable potatoes to be stored at colder temperatures and for longer periods without compromising fry quality and exacerbating acrylamide risk, extending the optimum storage window of potatoes and reducing the seasonality of acrylamide formation in crisps and other potato products. An enzyme involved in both processes is vacuolar invertase (VInv), which converts sucrose to glucose and fructose.

It is likely that further substantial reductions in acrylamide in potato products may not be achievable without the development of potato varieties with much lower concentrations of free asparagine and/or reducing sugars, together with improved stability of sugar concentrations during cold storage. GM potato varieties with these characteristics have already been developed and marketed in the USA by the Simplot Company of Boise, Idaho. The varieties are called Innate® and Innate® Generation 2. Both have reduced expression of a tuber-specific asparagine synthetase gene, *ASN1*, as well as reduced expression of two genes encoding enzymes of starch breakdown, phosphorylase L (*PhL*) and starch-associated R1 (*R1*), and a gene (*PPO5*) encoding polyphenol oxidase (PPO), an enzyme involved in bruising (Section 3.14), all as a result of RNAi. Innate® Generation 2 also has reduced expression of *VInv* and increased resistance to late blight through incorporation of the *Rpi-vnt1.1* gene from *S. venturii* (Section 3.12). The low concentration of free asparagine and reducing sugars in the tubers of Innate® Generation 2 is claimed to reduce acrylamide-forming potential by 90% compared with conventional potatoes. There is currently no prospect of these or any other genetically modified varieties being grown in Europe, highlighting how far behind crop biotechnology is in the UK and Europe compared with the USA, and how this is beginning to compromise efforts to improve food safety.

3.13.1. *Targeting vacuolar invertase by genome editing*

Improving potato storage characteristics and thereby reducing the potential of long-term storage to increase the risk of acrylamide formation in potato products has also been the aim of genome-editing programmes. The *VInv* gene, for example, has been targeted using TALENs

(Section 2.13.2) by teams from Cellectis Plant Sciences in the USA and France, using the popular French fry variety, Ranger Russet. Five lines were obtained with full *VInv* knockout, i.e. with mutations in genes within all four genomes (commercial potato being a tetraploid). Tubers from these lines had undetectable levels of reducing sugars, and crisps produced from them had much lower levels of acrylamide than crisps produced from control tubers.

As with the genome-edited lines with GBSS knockouts (Section 3.9.1), these lines have not yet been developed for commercial release.

3.14. Non-browning Apples and Mushrooms

The potato varieties Innate® and Innate® Generation 2 are best known for their reduced acrylamide-forming potential, but both also have reduced expression of a gene (*PPO5*) encoding PPO. PPO is one of the enzymes responsible for enzymatic or oxidative browning, the process by which phenols (chemical compounds comprising a hydroxyl group bonded to an aromatic hydrocarbon ring) that are naturally present in the food are converted to melanoidins and benzoquinone, creating a brown colour and affecting the flavour. The process requires oxygen (hence the term oxidative browning), and occurs rapidly not only in bruised potatoes but also in fruits such as apples when they are sliced (Figure 3.17). The rate of browning correlates with PPO activity, and *PPO* genes have therefore been targeted to reduce PPO activity and slow down the process.

As with the Innate® and Innate® Generation 2 potatoes, reduced PPO activity has been engineered into apples using RNAi. Arctic® Apples were developed by Okanagan Specialty Fruits Inc. of Canada and have been engineered with an RNAi gene comprising tandem regions of four apple *PPO* genes: (*PPO2*, *GPO3*, *APO5* and *pSR7*), under the control of a CaMV35S gene promoter (Section 2.11). Interestingly, they also contain a kanamycin resistance marker gene (Section 2.8), something that is considered to pose no risk to human health but which has generally been avoided in crop biotechnology since the 1990s due to the controversy that it provoked at that time. Arctic® Apples are aimed at meeting the increasing demand for fresh apple slices in fast food restaurants and single-serve packaging for retail. They were launched in the USA and Canada in 2017.

Figure 3.17.   Oxidative browning of a slice of apple, 10 minutes after slicing.

The non-browning trait has also been engineered into the common edible white mushroom, *Agaricus bisporus*. This is a fungus, not a plant, of course, but it is worth recording here because it was achieved by the genome-editing technique, CRISPR-Cas (Section 2.13.2). Furthermore, the United States Department of Agriculture's Animal and Plant Health Inspection Service (APHIS) has stated that it does not consider the genome-edited mushroom to be something that it must regulate; in other words, it does not consider the mushroom to be GM. This is discussed further in Section 4.6.

3.15. Drought, Heat and Cold Tolerance; Climate Change

The availability of water is a major determinant of plant production in many parts of the world, and shortages of water are recognised as major threats to food security. There is therefore an urgent need for new, high-yielding crop varieties with enhanced drought tolerance. However, developing high-yielding crops for water-limited environments is a major challenge, and progress has been limited by the complexity of the underlying traits, which are often determined by multiple genes and the seasonal and year-on-year variation of water availability. Drought is often, although not always, accompanied by high temperatures, and although these two stresses provoke different responses in plants the dividing line is often blurred as one stress exacerbates the effects of the other.

While both drought and heat stress are already problems in many parts of the world, the range over which they impact seriously on crop yields and the frequency with which they do it are both predicted to increase as a result of climate change. If the predictions of climate change are correct, plant breeders face the challenge of developing crop varieties for an environment that is going to change over the coming decades at unprecedented speed. The upper end of the range predicted by Global Climate Models in the IPCC 4th Assessment Report is an increase of 5°C in global mean temperature by the end of the century; this is a rate of change the like of which has not occurred in the last 50 million years. As well as the mean increase in temperature, there is predicted to be an increase in frequency of extreme weather events.

The potential impact of such events on food production can be seen in the effects of the severe Australian drought of the last decade, which reached crisis point in 2007–2008, and the Russian drought of 2010; both of these events were alluded to in Section 3.10. Both Australia and Russia are major wheat exporters in normal years, but in 2007–2008 exports of wheat from Australia fell to 20 million tonnes, while in July 2010 the Russian government banned wheat exports altogether to protect domestic supply. The price of wheat grain on the London International Financial Futures & Options Exchange (LIFFE) rose in 2008 to £198 (GBP) per tonne. They fell back to below 100 GBP per tonne in 2009 but rose to a new record of 200 GBP per tonne in December 2010. Less well known in the west but equally devastating was the drought suffered by southern China in 2013 (Figure 3.18), which raised concerns over future food security in the world's most populous country and played a big part in changing the thinking of the Chinese government on climate change.

In all three cases, these droughts were accompanied by high temperature, and both stresses would probably have affected crop yield. In wheat, for example, increased temperature shortens the growing period, reducing

Figure 3.18.   Total crop failure of upland rice growing in a field trial at Shanghai Academy of Agricultural Sciences, Shanghai, China during a period of extreme temperature and drought in August 2013. From: Halford, N. G., Curtis, T. Y., Chen, Z. and Huang, J. (2015) Effects of abiotic stress and crop management on cereal grain composition: implications for food quality and safety. *Journal of Experimental Botany* **66, 1145–1156.**

yield, while heat stress at flowering is especially damaging, resulting in much lower grain number and substantial yield losses. During the hot summer in northern Europe in 2003, for example, with maximum temperatures up to 38°C in the UK and 40°C in France, wheat production fell by about 20%.

Plants adopt several strategies to avoid the effects of drought. For example, if drought is most likely to occur in late summer (typical of the UK and northern Europe), they may avoid it by growing, flowering and setting seed before this time. Other avoidance strategies include the development of deeper and more extensive root systems, allowing the plant to obtain more water from the soil to survive dry periods. Plants have also evolved responses that enable them to survive even if they do become short of water. These are referred to as tolerance traits, and they differ from species to species and between different varieties, developmental stages, organs and tissue types. The plant hormone, abscisic acid (ABA), plays a key role, initiating a network of signalling pathways involving multiple protein kinases (enzymes that attach a phosphate group to another protein, affecting its activity) and transcription factors (proteins that regulate the expression of genes; Sections 1.3 and 1.4).

Transcription factors that are known to be involved in water stress responses include dehydration-responsive element binding protein (DREB)-1 and -2, ABA response element binding proteins (AREBPs), members of the zinc finger homeodomain (ZFHD)-1, myeloblastosis (MYB) and myelocytomatosis (MYC) families and the NAC family, which comprises no apical meristem (NAM), ATAF1 and 2, and cup-shaped cotyledon (CUC) transcription factors (these apparently odd names derive from the characteristics of mutant plants in which the transcription factor is not present). The action of AREBPs, DREB1, MYC and MYB requires ABA, while that of DREB2, ZFHD1 and NAC is ABA-independent. Over-expression of another transcription factor, plant nuclear factor-Y (NF-Y) has been shown to confer increased drought tolerance in maize in the field.

One of the areas of plant metabolism that is affected by these signalling pathways is carbohydrate metabolism, which plants manipulate to mitigate the effects of osmotic stress brought about by drought, for example by inter-converting insoluble starch with soluble sugars. Many of these responses are also seen in plants in response to cold, because cold, like drought, causes plant cells to suffer osmotic stress. Over-expression of protein kinases involved in ABA signalling has been shown to improve cold tolerance under laboratory conditions. Freezing is particularly dangerous to a plant because ice formation outside a plant cell prevents water uptake and causes an extreme osmotic stress leading to malfunctioning of cellular membranes, cellular dehydration and irrevocable damage. Plants prevent frost damage by modulating the concentrations of solutes and the lipid composition of their membranes. In 1991, a company called DNA Plant Technology attempted to improve the frost tolerance of tomato by introducing a gene encoding a so-called 'antifreeze' protein from a fish, the winter flounder. The project did not get beyond limited field trials and although similar experiments with other crop species have been performed it seems unlikely that such crops will reach the market in the near future. The prospect of crop plants being engineered with animal genes was a particularly emotive issue, and the 'fishy tomato' became an icon of the anti-GM campaign.

Many of the genes that are expressed in response to drought and cold stress are also expressed when heat stress is applied. In some cases, this may be misleading because elevated temperatures will cause water stress unless ambient humidity is adjusted to prevent it, and this sometimes obscures the fact that heat stress presents a plant with its own specific problems. As described above, high temperature causes wheat and other cereals to develop and mature more quickly; it also brings about an increase in respiration and an inhibition of photosynthesis. The latter is caused by a reduction in the activity of ribulose 1,5-bisphosphate carboxylase/oxygenase (Rubisco) and the efficiency of photosystem II. The reduction in Rubisco activity occurs because the enzyme responsible for maintaining its activity, Rubisco activase, is unstable at even moderately high temperatures. Genetic manipulation of Rubisco activase to improve its stability at high temperatures is therefore a potentially important target.

Another effect of high temperature is oxidative damage, and high activities of superoxide dismutase and catalase have been shown to be associated with good thermotolerance in wheat. Temperatures greater than 35°C during wheat grain development have also been shown to cause changes in the expression of different groups of seed storage proteins, with consequent effects on dough quality.

Transcription factors that are specifically associated with heat stress include heat shock factors (HSFs). The HSFs are quite a large family with at least 21 members, and their interaction appears to be complex. Nevertheless, simple over-expression experiments with HSFs have resulted in increased thermotolerance in transgenic plants. Heat stress can cause proteins, RNA and other molecules to fold incorrectly, affecting their assembly, translocation, turnover and activity. Heat-shock proteins (HSPs) and other so-called 'chaperones' that keep proteins and RNA in their correct conformation are expressed to mitigate this problem and these have therefore also attracted much attention. The expression of HSPs is under the control of HSFs.

All the major plant biotechnology companies claim to have drought-tolerant varieties in development, but Monsanto are the only company to have a product on the market at present. 'Drought-guard' maize was launched in 2012, and has been engineered to express bacterial RNA chaperones. The yield increases that are claimed are relatively modest, from 6.7 to 13.4% under drought conditions, but these varieties hopefully represent the start of a process of long-term improvement.

3.16. Salt Tolerance

Salt stress is linked with drought stress in both a plant physiological and a practical sense. In a plant physiological sense, both drought and salt impose an osmotic stress on the plant. In a practical sense, farmers respond to long-term drought by irrigating, usually with river water, and while river water is 'fresh' it does contain low concentrations of salt, with salt concentration rising nearer the sea. As the water evaporates from the land, it leaves the salt behind and the salt concentration in the soil gradually increases until it begins to affect plant growth and crop yield. This problem was recognised even in ancient times; the ancient Egyptians, for example, were aware that continuous irrigation of farmland with water from the Nile would affect soil fertility. It is now estimated that more than a third of all of the irrigated land in the world is affected by salinity.

High salt concentration in the soil prevents plants from taking up water. It also causes a build-up of Na^+ and Cl^- ions and skews the ratio of sodium and potassium ions. Plants respond by restricting the uptake of salt, adjusting their osmotic pressure by synthesising compatible solutes such as proline, glycine betaine and sugars, and sequestering salt in the cell vacuoles. They also neutralize ROS that are generated during the stress response.

Clearly, the development and use of crops that can tolerate high levels of salt in the soil would be part of a practical solution to the problem. As with drought and heat stress, however, progress has been hampered by the complexity of the underlying genetics. Consequently, only a few cultivars of the major crop species are registered as salt-tolerant. However, the existence of salt-tolerant plant species (called halophytes) and differences in salt tolerance between genotypes within salt-sensitive plant species (called glycophytes) indicate that there is a genetic basis to salt tolerance that could be exploited if it were understood.

As with drought and heat stress, plants respond to salt stress through the action of multiple signalling pathways involving protein kinases and transcription factors. Some of these pathways overlap or interact with the pathways induced by drought stress. There has been some success in improving salt tolerance, at least in the laboratory, through the manipulation of the genes involved in these pathways. However, more progress has been made through the introduction of novel genes or the manipulation of expression of native genes to impart ion homeostasis, osmotic regulation or antioxidant protection.

Na^+ extrusion from plant cells is powered by an H^+-ATPase which generates an electrochemical H^+ gradient that allows Na^+/H^+ antiporters (effectively molecular pumps) to pump Na^+ out of the cell. The over-expression of one of these antiporters, *SOS1,* in transgenic *Arabidopsis* has been shown to improve salt tolerance under laboratory conditions. There are also Na^+/H^+ antiporters that pump Na^+ out of the cytosol into the vacuole, where they do not cause damage. The over-expression of one of these, *NHX1,* in transgenic *Arabidopsis* and tomato has also been shown to result in plants that are able to grow in high salt concentrations. This work was pioneered by Eduardo Blumwald at the University of California.

Blumwald showed that tomato plants over-expressing *NHX1* could grow, flower and set fruit in the presence of 200 mM NaCl. Importantly, while the leaves accumulated high sodium concentrations, the fruits did not and were edible. The technology was subsequently applied to oilseed rape, with similar results. Leaves of the GM plants grown in the presence of 200 mM NaCl accumulated sodium to up to 6% of their dry weight, but the seed yields and oil quality were not affected. Other researchers have since had similar success with rice and maize.

A complementary strategy for improving salt tolerance concerns the synthesis and accumulation of a class of osmoprotective compounds known as compatible solutes. These are relatively small, organic metabolites, including some amino acids, such as proline, and their derivatives, such as glycine betaine (a small, trimethylated amino acid), polyols (alcohols with multiple hydroxyl groups) and sugars. The accumulation of these osmotically active compounds in the cytosol increases the cell's osmotic potential to balance the high concentration of Na^+ and Cl^- ions outside the cell. GM approaches have focused on increasing proline and glycine betaine synthesis. Increased proline accumulation has been induced by over-expression of genes encoding pyrroline-5-carboxylate (P5C) synthase (P5CS) and P5C reductase (P5CR), the two enzymes that catalyse the synthesis of proline from glutamic acid, and by reducing the expression of proline dehydrogenase, which is involved in proline breakdown. However, success has been limited by lack of understanding of the regulation of proline synthesis, metabolic limitations on levels to which proline can accumulate and a weak correlation between the concentration of proline and the salt tolerance that has been achieved.

Increasing glycine betaine accumulation has also proved difficult, hampered by problems of metabolic fluxes and cellular compartmentation of the substrate and product. Plants that naturally accumulate glycine betaine include spinach and sugar beet; in these species, synthesis of glycine betaine occurs in the chloroplast, catalysed by choline monooxygenase (CMO) and betaine aldehyde dehydrogenase (BADH). Bacteria have two pathways, one involving choline dehydrogenase (CDH) and BADH, the other a single enzyme, choline oxidase (COD, or COX). Engineering the second of these into GM plants is attractive because it involves only a single gene, and this has been done in *Arabidopsis*, *Brassica* and tobacco (as a model) using genes from *Arthrobacter globiformis* or *Arthrobacter panescens*. In some cases, the gene was first modified so that the encoded protein would be transported to the chloroplast, but improved salt tolerance was achieved in the GM plants with or without this modification. However, the degree of tolerance that was achieved was variable and glycine betaine could not be induced to accumulate to levels comparable with those in plants that accumulate it naturally. As with pro-line, a better understanding of the metabolic constraints on glycine betaine accumulation is required if more progress is to be made.

To conclude this section, it is clear that improved salt tolerance is possible; indeed, it has already been achieved. However, Blumwald reported his successful work on tomato in 1999 and there is still no commercial use of the technology. Part of the problem may be that the market is seen to be limited: while the area of land that is affected by salt globally may be large, it still represents a small proportion of the total agricultural land area, and most crops are not grown on salt-affected soils. Biotech companies may therefore be concerned that they will not get a return on the investment required for developing salt-tolerant varieties. If that is the case, we are unlikely to see such varieties on the market while the cost of development and regulatory compliance is so high.

3.17. Biopharming

Biopharming is the term applied to the use of GM plants to produce pharmaceuticals, vaccines or antibodies. It has also been applied to the production of proteins for industrial uses. Many vaccines and other pharmaceutical products are already produced in GM microbes, of course. Indeed, insulin was first produced in GM bacteria as early as 1981, enabling people with diabetes to be treated with human insulin instead of insulin from pigs or cattle. In 2007, a Canadian company, Sembiosys, a spin-out from the University of Calgary, announced that it had genetically modified safflower to produce insulin in its seeds. The attraction of using plants is that growing plants in the field would be relatively cheap compared with growing bacteria in sealed vats, and potentially would allow production on a much larger scale. Sembiosys pointed out that demand for insulin was increasing rapidly with the rise in cases of diabetes and the development of new delivery methods such as inhalation that require more insulin. However, if it was to be successful this product would have to compete with a well-established and widely accepted production system. Sembiosys ceased trading in 2012, and while it had established links with a number of partner companies in North America and China, the insulin-producing technology does not appear to have been taken to commercialisation.

The possibility of producing vaccines in GM plants has also caused great excitement, although progress has been slow since the idea was first put forward in the 1990s. It is estimated that there are 12 billion injections administered worldwide every year. Around 30% of these are not performed under sterile conditions. Furthermore, millions of people die every year from diseases that are preventable through vaccination. People may not have access to vaccines for one or more of several reasons. They may live in remote places that lack the infrastructure to deliver, store and administer the vaccine (transport, refrigeration, availability of needles and syringes). If they have to walk long distances to a clinic to have their children and themselves vaccinated, they may find it impossible to repeat the journey for a subsequent booster vaccination, so they are not protected adequately. There may not be sufficient numbers of medical professionals to administer a vaccination programme or the vaccine may just be too expensive. The production of vaccines in GM plants could potentially address some of these problems.

One of the first reports of successful vaccine production in a GM plant was for *Hepatitis B*. *Hepatitis B* causes acute and chronic disease of the liver and is associated with liver failure and liver cancer. It is a big killer throughout the developing world. The first *Hepatitis B* vaccine was a protein (the surface antigen) extracted from the blood of people infected with the disease and was produced in the 1970s. As HIV spread in the 1980s, this practice was viewed as too dangerous and the gene that encoded the protein was engineered into GM yeast to produce a safe vaccine. The vaccine has been made this way ever since and is extremely effective. Unfortunately, it is too expensive for many developing countries to afford, and even in the UK and other developed countries, the vaccine is only offered to those people considered to be at special risk of contracting the disease.

A group of researchers led by Hugh Mason at Cornell University genetically modified potato plants to make the *Hepatitis B* surface antigen in their tubers. The vaccine can be administered orally (patients eat a piece of the GM potato) and has been shown to work in animal and human trials. The original intention was that people would be able to grow plants producing the vaccine themselves. However, this would make the control of dosage impossible and is not now considered to be a realistic option.

Nevertheless, the advantages are clear: the potato can be stored at room temperature for a comparatively long time, and even moderate-scale production by agricultural standards could supply enough vaccine to meet global demand. Similar experiments have been performed to produce a vaccine for pathogenic *E. coli*, a bacterium that causes severe, sometimes fatal food poisoning, in potato and tobacco. However, the first human trials of this vaccine were completed in 1997, while results of a successful double-blind, placebo-controlled clinical trial of the orally administered *Hepatitis B* vaccine in potato were reported in 2005, yet there has been little sign of progress since.

These vaccines rely on a pathogen-derived protein or part of a protein to stimulate immunity. These so-called subunit vaccines cannot cause disease but may not always elicit the entire range of responses necessary to provide effective immunity. The *Hepatitis B* vaccine, for example, used the HbsAg surface antigen, while the pathogenic *E. coli* vaccine used the heat-labile toxin B subunit (LT-B). Other vaccines of this sort produced in plants include a *Rabies virus* antigen in tomato, a cholera antigen in tobacco and potato, the *Norwalk virus* capsid protein in tobacco and potato, and a *Human cytomegalovirus* antigen in tobacco.

The technology has also been applied to viruses that affect livestock. For example, the company ProdiGene, based in Texas, has produced an edible vaccine for *Transmissible gastroenteritis virus* (TGEV) in pigs, while the *Foot and mouth disease virus* (FMDV) antigen, VP1, has been expressed in *Arabidopsis*, potatoes and alfalfa.

A different approach to the production of vaccines in plants is the use of vectors based on plant viruses, and this has been used in particular for the production of vaccines for veterinary use. This approach has the advantages that very high levels of vaccine production can be achieved, viral genomes are small and relatively easy to manipulate, and infecting plants with a modified virus is easier than genetic modification of the plant. On the other hand, the production of the vaccine is transient, not heritable, there are limitations on the size and complexity of the DNA sequences that can be expressed in this way, and the genetic modification of viruses, rightly or wrongly, raises safety concerns.

One use of modified viruses has been to insert a DNA sequence encoding an antigenic peptide into the viral coat protein gene so that the peptide is expressed on the surface of assembled virus particles, often called chimaeric virus particles (CVPs). Antigenic proteins in multiple copies on the surface of a CVP are often much more effective at eliciting an immune response than when presented on their own. The modified viral particles are purified from the plant tissue for administration to animals. Alternatively, a gene is introduced into the viral genome so that it is expressed in infected cells, usually to make the protein on its own rather than fused with anything else. The protein may then be purified for administration, or animals could be fed plant material directly.

Plant viruses used in this way include *Cowpea mosaic virus* (CPMV), *Tobacco mosaic virus* (TMV), *Tomato bushy stunt virus* (TBSV), *Alfalfa mosaic virus* (AlMV), *Plum potyvirus* (PPV) and *Potato virus X* (PVX). Examples of promising results with the technique include protection of mink and dogs from canine parvovirus (CPV) by injection of a CPMV chimaera containing a short peptide from a CPV coat protein. Mice have also been immunised against *Mouse hepatitis virus* (MHV) by injection with chimaeric TMV particles containing a peptide from an MHV coat protein. In another experiment, mice were immunised against rabies with an AlMV chimaera.

The potential of vaccines produced in this way seems huge. However, there are problems to be overcome. These include scaling-up production, addressing safety concerns, negotiating regulatory hurdles, demonstrating that vaccines produced in plants are as effective as those produced by other means, and engaging pharmaceutical companies in plant vaccine development. Vaccines produced in plants may be administered by injection, but it was the notion of edible vaccines that caused much of the early excitement around plant-derived vaccines. It is now recognised that the notion of edible vaccines may be difficult to make a reality in practice because of problems with controlling dose and the requirement for an adjuvant (a protein that stimulates the immune response and increases the effectiveness of a vaccine). Possible adjuvants are themselves derived from pathogenic bacteria, and the most effective are toxic. There has been some success in recent years in using mutagenesis to produce non-toxic forms of some adjuvants and expressing these in plants, but the technology still requires some development and safety testing. Vaccines administered orally or nasally also present the difficulty of oral tolerance, which is a condition of unresponsiveness that can arise after a vaccine is administered in this way.

Plants have also been used to make antibodies. Antibodies bind target proteins with an extremely high degree of specificity and have a variety of uses in medicine, including disease diagnosis and the treatment of some cancers and autoimmune conditions. Their high degree of specificity means that they are also being developed for a number of uses outside medicine.

The production of antibodies in plants is not a trivial matter, partly because antibodies are complex proteins comprising multiple subunits. Full-length serum antibodies comprise two identical heavy (H) and two identical light (L) chains, and specificity is imparted by variable regions at one end of each chain. Secretory antibodies are even more complex, comprising dimers of the serum antibody structure plus a joining (J) chain and a secretory component (SC). This can be got around to some extent because less complex derivatives retain specificity. There are a number of different types of these derivatives, one example being single-chain Fv molecules (scFvs), which comprise the variable regions of the H and L chains of an antibody, held together by a flexible linker. However, antibodies are also glycosylated, in other words they have oligosaccharides attached to them, and the mechanisms of glycosylation in plants are different to those in animals. As a result, antibodies produced in normal plants are immunogenic in mammals. There has been progress in producing GM plants that glycosylate proteins in the mammalian manner, notably at the Universität für Bodenkultur in Vienna. These plants have been referred to as 'humanised' with respect to their glycosylation mechanisms and have been used to produce monoclonal antibodies that are equivalent to antibodies produced in mammalian cells.

One of the first and most impressive uses of plants for antibody production was the production in tobacco of a monoclonal antibody, Guys 13, which binds to the surface protein of *Streptococcus mutans*, the bacterium that causes tooth decay. Application of this monoclonal antibody to teeth prevents colonisation by *S. mutans*. Production of the antibody in plants was achieved by Julian Ma at Guys Hospital in London. The four subunits, H, L, J and SC, were each expressed in separate transgenic tobacco plants which were then crossed to produce a hybrid containing all four transgenes. The Guy's 13, so-called 'plantibody' technology, is licensed to Planet Biotechnology Inc. and has undergone Phase I and II clinical trials under the product name CaroRx™. More recently, Ma's team have been developing another monoclonal antibody made in tobacco, P2G12, for the treatment of HIV-1. P2G12 successfully completed a Phase I clinical trial in 2015.

Another project that looks promising is the expression in GM soybean of an antibody that recognises the *Herpes simplex virus* (HSV)-2 glyco-protein B. The plant-expressed antibody has been shown to protect mice against vaginal transmission of HSV-2. An example of the potential use of an antibody derivative is cFv 84.66, which recognises carcinoembryonic antigen (CEA), a protein that is present at elevated levels in people with colorectal, gastric, pancreatic, lung and breast cancer. The antibody could potentially be used to facilitate early diagnosis of these cancers, or to direct therapeutic drugs to cancer cells in the body (for example, it has been fused with interleukin-2 for this purpose). It has been produced in rice, wheat, tomato, pea and tobacco plants.

Antibodies for the treatment of patients with non-Hodgkin's lymphoma have been produced by Bayer through their subsidiary, Icon Genetics, and have undergone Phase I clinical trials. The antibodies are produced in tobacco (*Nicotiana tabacum*) using Bayer's magnICON® technology, in which a gene is introduced into a plant using an *Agrobacterium* species and spreads throughout the plant but does not insert into the plant's genome. Bayer therefore claim that the plants are not genetically modified, but the gene is expressed and the encoded protein is harvested from the plants' leaves.

Finally, ZMapp, which is made in the tobacco relative, *Nicotiana benthamiana*, and comprises three monoclonal antibodies against *Ebola virus*, has been shown to reverse Ebola disease in Rhesus macaques and was used to treat a small number of patients during the Ebola outbreak in Guinea, Liberia and Sierra Leone in 2013–2014. Two of the three antibodies in ZMapp were developed at the Public Health Agency of Canada's National Microbiology Laboratory, and the other at the U.S. Army Medical Research Institute of Infectious Diseases. ZMapp is being developed under licence by Mapp Biopharmaceutical, and clinical trials began in Liberia in 2015.

Clearly the production of vaccines and antibodies is difficult and complex. However, the term biopharming has also been applied to less challenging applications of GM, such as the production of enzymes for use in industry. One example of this is the production of trypsin, an animal protease (an enzyme that breaks down proteins), that has a variety of applications in biological research and is also used in the food industry, for example to pre-digest baby food. The company ProdiGene has genetically modified maize to produce trypsin, and the product, TrypZean, is already on the market for use in laboratory applications.

3.18. Removal of Allergens

Food allergy is an increasingly serious medical problem. The Food Standards Agency of the UK reported in 2000 that around 1.4–1.8% of the UK population and up to 8% of children in the UK suffer from some type of food allergy. It appears that there have been rapid increases in the number of incriminated foods and the frequency of severe reactions. At present, there is no specific treatment for food allergy, apart from dietary avoidance, although patients at risk of anaphylactic shock carry adrenaline in case of accidental exposure. Avoidance of allergens in processed foods can be difficult because their presence may not be obvious and products may not be labelled properly. The risk of accidental exposure, for example in a restaurant meal, is a constant anxiety that significantly affects quality of life.

Ironically, the possibility that some genes used in plant biotechnology could encode allergenic proteins has concerned consumers and regulatory authorities. This is considered in Chapters 4 and 5. The other side of the coin is that GM could be used to remove allergens from the food chain.

Well-known plant allergens include the 2S albumins, a group of storage proteins found in the seeds of legumes (peas and beans), crucifers (cabbages, turnips, swedes, mustards and radish) and many nuts. As well as acting as food stores for the seed, some inhibit digestive enzymes, perhaps protecting the seeds from being eaten, and some have anti-fungal properties.

Another group of proteins that includes several allergenic members are lipid transfer proteins, a family of small proteins present in seeds and other parts of a diverse range of plants, including fruits, oilseeds and cereals. They cause sunflower allergenicity in Southern European populations, resulting in severe symptoms, including anaphylaxis, albeit in a small number of sufferers. The function of lipid transfer proteins is not known, but they have been associated with plant defence.

Another family of proteins involved in plant defence are the PR (pathogen-related) proteins. These are made by plants in response to microbial infection. They include a number of allergenic proteins and are responsible for the allergenicity of chestnut, avocado, birch pollen, apples, cherries, celery and carrots.

Baker's asthma, a respiratory allergy caused by inhalation of wheat flour, is common in workers in the flour milling and baking industries. The allergenic proteins responsible appear to be small inhibitors of two digestive enzymes, α-amylase and trypsin. They are present in rye, barley, maize and rice as well as wheat.

The obvious target for biotechnologists is to remove these proteins from crop plants using gene suppression techniques (Section 2.12). These methods can lead to almost total suppression of gene activity and, while it might not be sufficient to make a food completely safe for someone who is allergenic to it, it could reduce the frequency and severity of reactions caused by accidental exposure. However, clearly this can only be done if the protein does not have an essential function in the plant.

There are currently only a few examples of experiments in which this has been tried. In one, the amounts of α-amylase inhibitors in rice were reduced substantially. There are at least 10 different genes encoding these proteins in rice, and to affect them all using conventional plant breeding or mutagenesis would be impossible. In another experiment, antisense technology was used to reduce the levels of the Lol p5 allergen in the pollen of ryegrass. The pollen of the GM ryegrass was shown to have reduced allergenicity.

These types of interventions obviously lend themselves to genome editing, which unlike GM gene suppression methods can be used to bring about a total knock-out of a target gene. However, I am unaware as yet of any reports of this being done successfully.

3.19. Conclusions

GM is now an established technique in plant breeding, complementing traditional methods and other new developments such as genomics-based accelerated breeding. Genome editing has arrived on the scene relatively recently, but the first edited crops are already in the field, and there is huge interest in the potential applications of genome editing in crop breeding. This chapter has described successful applications of GM in agriculture, some failures and some exciting prospects for the future. It is clear that GM is already bringing significant benefits to agriculture wherever GM crops are allowed to be grown. However, GM will only realise its full potential if the legislation covering the use of the technology does not make it too difficult or expensive to produce and market GM crops and consumers accept GM crop products. Similarly, genome editing will only attract investment if the regulations covering it are clear and workable. These issues are discussed in the next two chapters.

Chapter 4

Legislation Covering Genetically Modified (GM) Crops and Foods

4.1. Safety of Genetically Modified (GM) Plants Grown in Containment

Every country has developed rules and regulations covering the production and use of GM plants, but I will use the UK as an example. The question of GM crop and food safety was first considered in the UK in the early 1980s when the first GM plants were being produced. Responsibility for ensuring that the technology was developed safely was given to the Health and Safety Executive (HSE), a government agency responsible for the regulation of almost all of the risks to health and safety associated with the workplace in the UK. The HSE had already set up a committee, the Advisory Committee on Genetic Modification (ACGM), to control the use of GM microorganisms. The responsibility of this committee was extended to cover the production and use of all GM organisms (GMOs) in containment. The ACGM is now known as the Scientific Advisory Committee on Genetic Modification (Contained Use) (SACGM (CU)), and it provides technical and scientific advice on 'all aspects of the human and environmental risks of the contained use of GM organisms'.

No organisation in the UK can produce or hold GMOs without the permission of the HSE. The HSE ensures that any organisation that proposes working on GMOs has the facilities required and has staff trained and experienced in the handling and disposal of the organisms and contaminated waste. Different levels of containment are required for different GMOs, depending on the risk that they represent.

An organism is said to be in containment if physical, chemical or biological barriers are used to limit contact between it and other organisms or the environment. Biological containment is often achieved with bacteria by using strains that are genetically disabled and therefore unable to survive outside the laboratory. It can be applied to plants by using species that do not survive in the local environment and do not cross with native wild plant species (wheat, for example, does not survive outside agriculture and has no wild relatives in the UK), or by adopting measures such as the removal of flowers before pollen is shed. An example of a chemical barrier to the spread of a GM plant is the use of an herbicide, although it would be more likely to be applied in the post-harvest treatment of a field test site than in strict containment conditions. Physical containment involves the use of specially designed laboratories, plant growth rooms and glasshouses. There are four categories of containment for GM microorganisms and two for GM plants and animals (Table 4.1). In the case of GM plants, the legislation only covers their use in the laboratory and glasshouse; field releases are covered separately.

Even for GMOs that are considered to be of no risk to human health or the environment, the laboratory in which they are held and used must satisfy basic standards. For example, the laboratory must be easy to clean, benchtops and floor must be sealed, and if the laboratory is mechanically ventilated the air flow must be inwards. Hand-washing facilities must be provided next to the exit and basic protective clothing, such as a lab-coat and disposable gloves, must be worn and removed before leaving the laboratory. Access to the laboratory must be restricted. Any procedures that produce aerosols must be done in a safety cabinet, effective disinfectants must be available next to every sink, spillages must be dealt with immediately and recorded, benchtops must be cleaned after use and a good general standard of cleanliness must be maintained. All contaminated glassware must be stored safely and sterilised after use, and all

Table 4.1. Risk categories of GM microorganisms, plants and animals kept in containment in the UK.

	GM microorganisms				GM plants and animals	
	Class 1	Class 2	Class 3	Class 4	A	B
Risk to human health or environment	None or negligible	Low	Medium	High	Less risky than non-GM parent	More risky than non-GM parent
Notification requirement	HSE must be notified that GM work is to be undertaken	HSE must be notified of each activity in these categories			HSE must be notified that GM work is to be undertaken	HSE must be notified of each activity
Containment and safety measures	Increasing requirement for measures to contain organism and protect handler				Increasing requirement for containment measures	

contaminated waste must be stored safely in a designated bin and auto-claved (sterilised by heat and pressure) before disposal.

At the heart of the safe handling of GMOs is expert risk assessment. All projects involving GMOs have to be risk-assessed by the project leader, and the risk assessment must be considered by an organisation's internal Genetic Modification Safety Committee or Biological Safety Committee. The project leader is required to give information on the experience level of staff who will work on the project and the training that they have received. For a project concerning GM plants, information has to be given on the host plant and the gene or genes being inserted. To assess any risk to human health, the project leader has to consider any possible induction of or increase in toxicity and/or allergenicity compared with the parent plant and the risk of accumulation of toxicity through food chains.

The project leader also has to assess any risk posed by the proposed GM plants to the environment, particularly the potential of the plants to be more 'weedy' than the parent plant. This assessment includes factors such as colonisation ability, seed dispersal mechanisms, resistance to control measures such as herbicides, increased toxicity to insects and other grazers and any other possible change in the plants' interaction with their environment. The potential for and consequences of the sexual transfer of nucleic acids between the GM plants and other plants of the same species or a compatible species has to be considered, particularly if the plants have the ability to transfer novel genetic material to UK plant species. The risk and consequences of horizontal gene transfer to unrelated species, for example by a virus, bacterium or other vector, also has to be taken into account. Finally, an assessment has to be made of the potential of the GM plants to cause harm to animals or beneficial microorganisms.

This risk assessment has to be completed before the experiment begins and despite the fact that the plants will be kept in containment in specially designed glasshouses or controlled environment rooms. These have filtered negative air pressure ventilation, sealed drains and a chlorination treatment system for drainage water to ensure that no viable plant material escapes into the environment.

The proper risk assessment of experiments involving GMOs and their safe handling and containment are legal requirements and failure to observe the regulations could result in prosecution. It would certainly result in serious cases in the loss of an institution's licence to work with GMOs. All institutions that work with GMOs are inspected regularly by the HSE.

4.2. Safety of Field Releases of GM Plants

In the late 1980s, it seemed that the growing of GM plants in the field in the UK was likely to become widespread and the government decided that it would have to be controlled. Regulations covering the field release of GM plants and animals were written into the Environmental Protection Act of 1990 and a statutory advisory committee was set up to provide advice to government regarding the release and marketing of GM organisms. This committee was called the Advisory Committee for Releases into the Environment (ACRE). The regulations were updated through the GMO Deliberate Release Regulations 2002, which implemented the European Union's directive on the deliberate release of GMOs: Directive 2001/18/EC.

ACRE advises the Secretary of State for Environment, Food and Rural Affairs, Scottish Ministers, the Welsh Assembly Secretaries and the Department of the Environment in Northern Ireland. It also advises the HSE on human health aspects of releasing GMOs. ACRE is made up predominantly of academics and, between them, members have expertise in agronomy, biodiversity, ecology, entomology, microbiology, plant molecular biology, plant breeding, plant physiology and farming practice.

Anyone who wants to undertake a field release of a GM plant in the UK has to obtain permission from the government's Department for Environment, Food and Rural Affairs (DEFRA). A detailed risk assessment of the plants must be undertaken and considered by ACRE, who then advise DEFRA on whether or not to allow the release to go ahead. More information will be available for this risk assessment than the one described for a proposal to make a set of GM plants and grow them in containment because the plants will already have been made and studied. A typical risk assessment will include:

- Information on the host plant species:
  — The full name of the plant species and the breeding line used.

- — Details of the sexual reproduction of the plant, generation time and the sexual compatibility of the plant with other cultivated or wild plant species.
- — Information concerning the survivability and dissemination of the plant and the geographical distribution of the species.
- A description of the methods used for the genetic modification and the nature and source of the vector used to modify the plant.
- Information on the nature of the genetic modification:
  - — Details of the novel genes introduced into the plant, including size, intended function and the organisms from which they originated.
  - — Information on when and where in the plant the novel gene or genes is/are active and the methods used for finding out.
  - — Information on the location of the inserted novel DNA in the plant cells and the number of copies of the novel gene or genes that are present.
  - — The size and function of any region of the host plant genome that has been deleted as a result of the genetic modification.
  - — An analysis of the genetic stability of the novel gene or genes.
- An assessment of the GM plants:
  - — A description of the trait or traits and characteristics of the GM plant which have been introduced or changed.
  - — An assessment of any differences between the GM plant and its parent in respect of methods and rates of reproduction, dissemination and survivability.
  - — A description of detection and identification techniques for the GM plants.
- An assessment of potential risks posed to health and/or the environment:
  - — An assessment of any potential toxic effects on humans, animals and other organisms.
  - — An assessment of the likelihood of the GM plant becoming more persistent than the recipient or parental plants in agricultural habitats or more invasive in natural habitats.

— A description of the mechanism of interaction between the GM plants and target organisms (for example, if the plants have been engineered to be resistant to insects) and any potentially significant interactions with non-target organisms.
— An assessment of the potential environmental impact of the interaction between the GM plant and target or non-target organisms.
— An assessment of the potential for transfer of genetic material from the GM plants to other organisms.
— Any selective advantage or disadvantage conferred to other species that may result from genetic transfer from the GM plant.
— Information about previous releases of the GM plants.

• Information on the release site:

— The location and size of the release site or sites.
— A description of the release site ecosystem, including climate, flora and fauna.
— Details of any sexually compatible wild relatives or cultivated plant species present at the release sites.
— The proximity of the release sites to officially recognised protected areas that may be affected.

• A description of the management of the field trial:

— The purpose of the trial.
— The foreseen dates and duration of the trial.
— The method by which the GM plants are to be released.
— The method for preparing and managing the release site, including cultivation practices and harvesting methods.
— The approximate number of GM plants (or plants per m^2) to be released.

• A description of containment measures to be adopted during and after the field trial:

— A description of any precautions to maintain the GM plant at a distance from sexually compatible plant species and to minimise or prevent pollen or seed dispersal.
— A description of the methods for post-release treatment of the site. These are likely to include ploughing of the site, irrigation to

encourage germination of any seed in the soil, the removal of any plants that sprout by spraying with an appropriate total herbicide and a period of one to 2 years when the site is kept fallow and monitored.

— A description of how the GM plant material will be disposed of.
— Details of how the site is to be monitored after the trial is over.
— A description of emergency plans in the event that an undesirable effect occurs during the trial or that the plants spread.
— An assessment of the likelihood and consequences of theft of GM material from the trial, vandalism of the trial or accidental movement of GM material off the trial site by any means (e.g. on field machinery).

Clearly, providing all of this information is a lengthy and expensive business. On top of that, the cost of a new application is £5,000, with each actual release costing an additional £850 per year; not insignificant in the context of research budgets. Applicants have to wait at least 90 days for a decision, with the clock stopped if the applicant is asked for further details until the information is provided. This delay can be a problem because, obviously, seed has to be sown at a particular time of year. Applicants are also required to provide the map reference of the trial site and place this in the public domain through an advert in the local press notifying the public of the trial. As a result, many field trials in the UK have been vandalised and most companies and institutions have decided that it is not worth attempting to run them. Indeed, there were no GM crop field trials at all in the UK between 2004 and 2009, and the only trial being run in the UK at present is at my own institution, Rothamsted Research.

Rothamsted Research had ceased running GM field trials in the early 2000s as a result of repeated vandalism at the site. It re-started trials in 2012 with a field trial of GM wheat that emitted an insect alarm pheromone to deter aphids. The lengths that had to be gone to in order to run the trial are illustrated in Figure 4.1. Small plots of GM wheat had to be surrounded by barley as a 'separator', then non-GM wheat as a pollen barrier. The trial site had to be surrounded with a security fence and ditch (to prevent ramming of the site with a vehicle), and protected by

GM wheat plots

Separator (barley)

Pollen barrier (non-GM wheat)

Security fencing and ditch

Figure 4.1.   Field trial of GM wheat, Rothamsted Research, UK, 2012.

round-the-clock security patrols, increasing the cost of the project from £732,000 to approximately £2.5 million. The trial started in 2012, and as the only GM field trial in the UK that year, and the first involving wheat for over a decade, it attracted considerable notoriety and a small demonstration at the site. Nevertheless, the trial was completed successfully and repeated in 2013 and 2014.

The GM wheat plants had been shown to repel aphids and attract beneficial predators in glasshouse experiments, but this was not replicated in the field, with no improvement to yields, reduction in aphids or increase in aphid predators. This was disappointing, but, of course, the point of field trials is to determine whether a result in a glasshouse can be transferred to the setting where it really matters.

The GM wheat field trial has been followed by field trials of GM *Camelina* producing omega-3 long-chain polyunsaturated fatty acids (omega-3 LC-PUFAs) (Section 3.8.3.1). The Rothamsted Research field trial remains the only GM field trial being run in the UK. In contrast, tens of thousands of field trials of GM plants have been performed in the USA on species from soybean to walnut. The vast majority of these have caused no problems, and the American authorities have made the application procedure much easier as a result. However, they are discovering that with some GM plant experiments, such as those involving plants that make pharmaceutical products, they have to be more careful (Section 5.17). Nevertheless, the fact that field experiments are so much more difficult in the UK and Europe than in the USA does put European researchers at a competitive disadvantage. Despite the fact that no problems have arisen from the field experiments of GM plants that have taken place in Europe, there is no indication that the regulations will be relaxed in the foreseeable future.

4.3. Safety of GM Foods

The UK government can take advice on the safety of foods derived from GM crops from another committee, the Advisory Committee on Novel Foods and Processes (ACNFP). ACNFP was established in 1988 to advise the responsible authorities in the UK on any matters relating to novel foods and novel food processes. It comprises 15 members, most of whom are academics, with, between them, expertise in allergenicity, genetics, immunology, microbiology, nutrition and toxicology. It comes under the jurisdiction of the Food Standards Agency (FSA).

The assessment made by the ACNFP essentially follows guidelines endorsed by the World Health Organization (WHO), and the same guidelines are followed by almost all regulatory authorities around the world. At the heart of the process is the concept of substantial equivalence. This concept is often attacked by anti-GM campaigners, who claim that it merely means showing that a GM plant, or food derived from it, is roughly the same as its non-GM counterpart. In reality, substantial equivalence involves a comprehensive biochemical and molecular comparison of a GM food and its conventional equivalent and a detailed analysis of any differences.

The fact is that very few foods consumed today have been subject to any toxicological studies, yet they are generally accepted as being safe to eat. The difficulties of applying traditional toxicological testing and risk assessment procedures to whole foods, GM or otherwise, makes it pretty well impossible to establish absolute safety. The aim of the substantial equivalence approach, therefore, is to consider whether the GM food is as safe as its traditional counterpart, where such a counterpart exists.

The process begins with a comparison between the GM food and its closest traditional counterpart in order to identify any intended and unintended differences. These differences then become the focus of the safety assessment and, if necessary, further investigation. Factors taken into account in the safety assessment include:

- The identity and source of novel genes (in particular is the source a well-characterised food source or is it entirely new to the food chain).
- Composition of the plant and/or food derived from it compared with its traditional counterpart.
- Effects of processing/cooking.
- The methods used to make the GM plant.
- The stability and potential for transfer of the novel gene or genes.
- The nature of the protein encoded by the novel gene or genes.
- Potential changes in function of novel genes and proteins.
- Potential toxicity of novel proteins.
- Potential allergenicity of novel proteins.
- Possible secondary effects resulting from expression of the novel gene or genes, for example, by disruption of a gene in the host plant, knockon effects on metabolic pathways and changes in the production

of nutrients, anti-nutrients, toxins, allergens and physiologically active substances.

* Potential intake and dietary impact of the introduction of the GM food.

The technology available to undertake these sorts of studies has moved on tremendously in the last few years. The activity of thousands, sometimes tens of thousands, of genes can be determined in a single experiment. Similarly, the amounts of thousands of proteins and metabolites present in a plant can be measured. These techniques of transcriptomics, proteomics and metabolomics are revolutionising safety testing. However, it is possible, likely even, that changes in gene expression, protein synthesis and metabolite profiles would be found in new varieties of crops produced by any of the methods used in plant breeding, whether it be crossing, mutagenesis, genetic modification, genome editing or anything else. The difficulty will be in identifying those changes that are significant and making rational decisions about them.

Some work has already been done to compare the effects of GM and other factors on wheat seed composition at the metabolomic centre at Rothamsted Research, UK (MeTRO). The substantial equivalence of three GM wheat lines and their non-GM parents was examined using metabolite profiling of samples grown in field trials on sites near Bristol, in the west of the country, and at Rothamsted itself in the east, over 3 years. The conclusions of the study were that the differences in grain composition caused by site and year of cultivation were greater than those caused by genetic modification. A similar study of global gene expression (transcriptomics) in transgenic and conventionally bred wheat lines showed that the differences in gene expression between conventionally bred genotypes were much larger than those between GM and non-GM genotypes.

4.4. European Union Regulations

The European Union's directive, GM Food and Feed Regulation (EC) No. 1829/2003, was adopted in 2004, bringing the regulation of GM crop use and release under the control of the European Commission. The EU recognises two different types of field release of GM crops, one for research purposes only (a Part B release) and one for commercial release (a Part C release). Consent for a Part C release may be granted for cultivation, food and feed use, or for food and feed use alone (i.e. for import and consumption in the European Union but not for cultivation). Permission for a Part B release can be granted by an individual Member State, as described above for the UK. However, applications for a Part C release anywhere in the European Union have to be approved by the European Commission. That means applications concerning any commercial use of GM crops for cultivation, food or feed use in the European Union.

Each new application is assessed by the European Food Safety Authority (EFSA). If EFSA approves the application, it is voted on by a working group with representatives from all Member States (currently 28). The UK representative may take advice from ACRE, ACNFP and another UK advisory committee, the Advisory Committee for Animal Feedingstuffs (ACAF), as well as considering the assessment of EFSA. The European Commission has adopted a complicated system of Qualified Majority Voting (QMV), in which approval requires 55% of Member States to vote in favour (in practice this means 16 out of 28), and the supporting Member States to represent at least 65% of the total EU population. It is therefore relatively easy for a minority of countries to block an application: a blocking minority can be made up of only four council members if they represent more than 35% of the EU population. If that happens, the Council of Europe and even the Council of Ministers may become involved in a series of futile cycles.

BASF's Amflora potato variety (Section 3.9) was bogged down in this process for 13 years, which is a farcical situation. Indeed, the European Commission failed to approve a single application between 1998 and 2004 because six member states, France, Italy, Denmark, Greece, Austria and Luxembourg, managed to block every one. This *de facto* moratorium undoubtedly damaged trade relations and stymied the development of the plant biotechnology industry in Europe. The impasse was finally broken in May 2004 when the European Commission authorised Syngenta Bt-11 sweetcorn, an insect-resistant and gluphosinate-tolerant variety, for import and food use. Later in 2004, it extended the consent for MON810, an insect-resistant variety that had been authorised for food use throughout the EU and for cultivation in France and Spain since 1998. The extension allowed for cultivation of this variety, which has been quite successful in Spain (Section 3.6.2), throughout the EU. Note that individual Member States still imposed restrictions, however, in contravention of EU regulations. France, for example, banned the cultivation of this variety in 2008 after it had appeared to be becoming popular with farmers in 2006 and 2007.

The logjam in applications started to shift after 2004, but progress was still slow, and Austria, France, Germany, Luxembourg and Greece retained their own national bans in defiance of the European Commission. Frustrated with the lack of progress, the USA, Australia, Argentina, Brazil, Canada, India, Mexico and, surprisingly, New Zealand, complained to the World Trade Organisation (WTO) and, in 2006, the WTO ruled that the European Union's position was illegal and also criticised the bans imposed by individual Member States.

It remains extremely difficult to obtain Part C consent for cultivation and marketing of GM crops and their products in the European Union, and the area of GM crops in Europe is still in the hundreds of thousands of hectares (Figure 3.6) in contrast to the tens of millions of hectares in the Americas and Asia. It is easier, although still problematic, to gain approval for food and feed use alone without seeking approval for cultivation. As a result, efforts to develop new varieties for cultivation in Europe have all but been abandoned by biotech companies. Instead, companies are focussing on obtaining permission for food and feed use so that their potential customers elsewhere in the world can be reassured that the European market is open to their products. This means that European farmers are competing with GM crops but are unable to use them. Indeed, only two varieties of GM crops are currently approved for cultivation in Europe: MON810, marketed by Monsanto, and the Amflora potato developed by BASF (Section 3.9). Of these, the Amflora potato variety is no longer available, meaning that MON810 is effectively the only GM crop variety available to European farmers. Note that the Bt trait has been engineered or crossed into many varieties of maize, but the Part C consent for MON810 applies only to that variety, and any other Bt maize variety would have to go through the whole approval process before it could be marketed for cultivation.

4.4.1. *Devolution of decision-making to Member States*

Recognising that getting agreement from all European Union Member States was proving too difficult, the European Commission proposed that some of the decision-making on GM crop and food issues should be devolved back down to national Member State governments, in an apparent attempt to reconcile the implacable positions of pro- and anti-GM Member States. In March 2015, after years of dialogue, the European Union approved the Commission proposals for a new system of GM crop authorisations, based on a two-step process. Step 1: Member States may request, via the Commission, that a company applying for approval of a GM crop adjust the geographical scope of its application for authorisation so that it does not include a particular Member State or region. The usual risk assessment in which EFSA makes an independent scientific assessment will then follow for a defined geographical area before a GM crop is finally approved. Step 2: If the applicant refuses to adjust the geographical scope of its authorisation, Member States will be allowed to proceed with national bans.

Whatever this was supposed to achieve, in simple terms it allows Member States to opt out of allowing cultivation or import of a GM crop that has been approved at EU level, without giving a reason, whereas under the previous system countries could only legally opt out if they had a solid scientific basis for doing so. Within weeks of the new legislation being adopted, 15 Member States told the European Commission that they would send territorial exclusion requests to Monsanto, Dow, Syngenta, Pioneer and other biotech companies. These countries were Austria, Bulgaria, Croatia, Cyprus, Denmark, France, Germany, Greece, Hungary, Italy, Latvia, Lithuania, the Netherlands, Poland and Slovenia. In addition, Scotland, Northern Ireland and Wales said they would opt out on a regional basis, as did Wallonia (the French-speaking region of southern Belgium).

It is highly unlikely that any biotech company will attempt to develop a GM crop for cultivation in Europe while this system operates. This is not only affecting the competitiveness of European agriculture, but also stymieing efforts to improve food nutritional value and safety

(Sections 3.8.3, 3.11 and 3.13). Furthermore, because the European Union is the biggest commodity market in the world, European regulations and attitudes are a wet blanket for countries that supply it, with many African countries that would benefit hugely from GM crops afraid to adopt them for fear of losing access to the European market.

4.4.2. *Risk assessment of GM crops with deliberately altered composition*

Another challenge facing the European Commission is the prospect of having to adapt its cumbersome processes to the risk assessment of GM crops, foods and feeds with deliberately altered composition (as opposed to those carrying herbicide tolerance or insect resistance traits, for example, in which any compositional changes have arisen inadvertently). In effect, this means second-generation GM crops with output/quality traits (Section 3.3). As the number of GM varieties around the world with these types of traits increases (Sections 3.4, 3.8, 3.9, 3.10.1, 3.11, 3.13, 3.17 and 3.18), the European Commission believes that it is only a matter of time before it will have to risk assess them for import for food and feed use.

As described above, the Commission's current risk assessment strategy, as applied by EFSA's GMO Panel, is based on the principle of substantial equivalence, in which GM plants and their derived food and feeds are compared with a conventional, non-GM comparator, the comparator preferably being a plant with a history of safe use as food. The development of GM crop varieties that have received substantial, deliberate modifications to their composition, metabolism or physiology raises the question of whether the comparative approach would be fully applicable in some cases. Indeed, EFSA guidance on the risk assessment of food and feed from GM plants states that: 'Where no comparator can be identified, a comparative risk assessment cannot be made and a comprehensive safety and nutritional assessment of the GM plant and derived food and feed itself should be carried out. This would, for instance, be the case where the food and/or feed derived from a GM plant is not closely related to a food and/or a feed with a history of safe use, or where a specific trait or specific traits are introduced with the intention of changing significantly the composition of the plant'.

An example of a GM crop with altered composition that fell foul of the current EFSA risk assessment process is the 'Amflora' potato variety (Section 3.9). 'Amflora' was developed by BASF and contained starch composed almost entirely of amylopectin as a result of reduced activity of a granule-bound starch synthase. The aim was to produce starch as an industrial raw material; however, the application for permission to market the variety took 13 years to pass through the EU approval process. It was finally cleared for cultivation in 2010, but BASF announced in 2012 that it was withdrawing from plant biotechnology in Europe altogether, and although 'Amflora' is licensed for cultivation it is currently unavailable. The fiasco led to accusations that the licensing process was dysfunctional, and highlighted the need for robust, workable processes for the safety assessment of GM crops with altered composition.

I was involved in the preparation of a report on this issue for the European Commission in 2013 (EFSA supporting publication 2013:EN-480). The conclusions of the report were that whereas the Commission foresaw the need for 'comprehensive' risk assessment strategies to be developed that did not rely on the comparative approach to assess safety, comparative approaches based on the concept of substantial equivalence were the basis of all current risk assessment strategies across the world, for second-generation as well as first-generation GM crops. In other words, a comprehensive assessment of GM plants and derived food and feed without the use of a comparator is not an approach used by international risk assessment bodies or proposed by the scientific literature. Indeed, second generation GM varieties that had already been approved in the USA and other countries outside the EU had been risk assessed using the same framework as those with first generation input traits, with no specific modifications to risk assessment criteria.

Nevertheless, the report conceded that new approaches would need to be developed to assess novel GM crops containing constituents that are not normally found in that species, with GM oilseeds containing LC-PUFAs that are otherwise found only in fish oils being an obvious example (Section 3.8.3.1). The concept of 'history of safe use', which is already a valuable tool in risk assessment, would be particularly useful in such circumstances: If an applicant could demonstrate that the novel constituent in the GM plant had a 'history of safe use' in food or feed, even if it had previously come from a different source, this would give confidence in its safety. However, while this concept was and is being used by many international risk assessment bodies, it has not been defined; in particular, there is no standard requirement for time period or level of consumption. In the report, the definition used by Health Canada was supported, but even this uses vague terms: 'the safety of the expressed trait in food or feed is confirmed from experience of use and continued use in the normal diet of a large part of the population of a country or in farmed/domestic animals for a number of generations, consumed at levels foreseen to be similar to its use in the GM plant'.

It may sometimes be more useful to compare a food product derived from plants carrying a novel trait to another food product that is more similar to it than its conventional counterpart. In the case of GM oils containing LC-PUFAs, this could be fish oil. However, GM plants that have been engineered to synthesise LC-PUFAs do not make true fish oils; they make an oil with a unique combination of some of the fatty acids that would normally be present in that plant species and some additional LC-PUFAs. So, although informative, such a comparison could only be part of the process.

Another consideration would be whether the novel GM crop would change consumption patterns and therefore exposure. This could be difficult to predict because it would be affected by consumer preferences, which could change over time, and the trend by food manufacturers to include functional ingredients in multiple products. It would be particularly important for food constituents for which there is an upper limit to the recommended dietary intake (Vitamin A, for example (Section 3.11)). It would also be important that consumers were given enough information about the difference between the GM and non-GM products to make informed choices over consumption, an issue already discussed in Section 3.8.2.

Whatever system is put in place for risk assessing and licensing second generation GM traits, it must be robust, transparent and efficient.

4.5. Safety Assessment of GM Crops in the USA

New GM crop varieties have to undergo field trials in the USA in the same way that they do in Europe. However, much less detail is required in the risk assessment of the variety before the trial can go ahead. In fact, the procedure has been relaxed as more field trials have been undertaken and no problems have ensued. The total number of field trials of GM crops that have been run in the USA now runs into the tens of thousands, covering a variety of crop species and traits. However, the process may be tightened up again as applications are made to test crops that produce pharmaceuticals or have other non-food uses.

Test and commercial releases of GM crops in the USA are controlled by the Animal and Plant Health Inspection Service (APHIS) within the United States Department of Agriculture (USDA). Commercial use of GM crops that produce their own insecticide, such as Bt, is controlled by the Environmental Protection Agency (EPA). The safety of foods derived from GM crops is assessed by the Food and Drug Administration (FDA). In fact, there is no legal requirement for companies to seek the approval of the FDA but all of them do. Companies ask the FDA for an Advisory Opinion either on a specific characteristic of their product or on its suitability as a food. Liability for problems that arise after release rests with the company that markets the crop.

The first GM plant product to come onto the market in the USA was Calgene's Flavr Savr tomato (Section 3.4). It underwent more than 4 years of comprehensive pre-market tests, the results of which were submitted to both the FDA and the USDA and were published for public comment. Calgene requested two separate Advisory Opinions from the FDA, one on the use of the kanamycin resistance marker gene, the other on the status of the Flavr Savr tomato as a food.

The FDA issued a preliminary report that all relevant safety questions about the Flavr Savr tomato had been resolved. This was ratified by its Food Advisory Committee in a public meeting in April 1994. The FDA announced its findings that the Flavr Savr tomato was as safe as tomatoes bred by conventional means in May 1994, and Calgene began to market the new product shortly after.

A third agency, the EPA, became involved in the regulatory process for GM crops when Bt crops were developed, and still is involved if the GM crop being assessed has increased resistance to insect pests (Section 3.6). The EPA insisted that it should have a role in regulating Bt crops when they were first introduced in the 1990s because it was responsible for regulating the use of Bt when applied as a conventional pesticide (i.e. sprayed onto crops in the form of powders, granules, or aqueous and oil-based liquids). It was EPA's modelling that suggested that unrestricted use of Bt crops could lead to the development of resistance in its target species, making not only the new Bt varieties useless but also the Bt pesticides that were relied upon by organic and salad farmers. The EPA forced farmers planting Bt crops to plant a certain proportion of non-Bt crops as well, providing refuges for non-resistant insects, which meant that any insects developing resistance would not be at a selective advantage (discussed in more detail in Section 5.11). Bt crops have now been used in the USA for more than 20 years without resistance developing in the target insect species.

4.6. Regulating Genome Editing

The development of genome editing has brought home to many regulatory authorities that the regulations that were drawn up in the 1990s need to be updated in light of the technological advances. Needless to say, the European Commission is in the most difficult position because its regulations are so cumbersome anyway, and because it considers the political ramifications and criticism from NGOs ahead of the scientific case.

Chemical and radiation mutagenesis have been used in plant breeding since the 1950s (Section 1.8.2) and are exempt from the regulations covering GM crops, which were, of course, drawn up decades later (the mutagenesis exemption). These methods introduce thousands of mutations into a plant's genome, totally randomly, but new varieties produced this way can be released onto the market without any more scrutiny or regulation than any other variety. Mutations also occur naturally, and in the past the products of these naturally occurring mutations have been adopted for food use (Section 1.7). In their most common usage, genome-editing techniques introduce one mutation into a single target gene, and any transgene used in the process, if there is one, can be eliminated. From a scientific point of view, therefore, there can be no justification for treating genome-edited plants differently to plants produced by chemical or radiation mutagenesis, and genome-edited plants should not be subject to the regulations applied to GM plants.

That certainly appears to be the view of USA regulatory authorities: The USDA has stated that it will not regulate a mushroom genetically modified with the gene-editing tool CRISPR-Cas9 (Section 3.14). Nevertheless, the USA is reviewing its regulations on agricultural biotechnology in the light of developments in genome editing, and that process is ongoing at the time of writing. In September 2016, the Office of Science and Technology Policy issued a National Strategy for Modernising the Regulatory System for Biotechnology Products, setting out a 'vision' for updating the federal regulatory system so that it would be able to assess future products of biotechnology. It followed that up in January 2017 by issuing an 'Update to the Coordinated Framework for the Regulation of Biotechnology', in which it clarified the roles and responsibilities of the FDA, EPA and USDA/APHIS. The FDA has also opened a public consultation on the regulation of foods from plants produced using genome editing.

Argentina and Canada have also developed policies for genome-edited plants, essentially saying that new varieties will be assessed on a case-by-case basis but that if there is no new combination of genetic material and no transgenes have been used, the product is non-GM. That includes plants in which a transgene was used in the development of a new variety, but has been removed before commercialisation (the case for most plants produced using targeted nuclease techniques, such as TALENs and CRISPR-Cas (Section 2.13.2).

In contrast, the European Commission has ducked the issue, deferring a decision on the regulation of genome-edited plants to an unspecified future date. That is despite the fact that an expert on EFSA's GM panel has stated that 'There are some applications of these [genome editing] techniques (i.e. where no recombinant DNA remains in the commercialised plant and where the "edit" is indistinguishable from a natural mutation) that seem not to be a GMO as currently defined'. One EU Member State, Sweden, has gone it alone: in December 2015, the Swedish Board of Agriculture, in response to formally submitted questions from Swedish scientists, issued a statement that its interpretation was that plants in which the genome has been edited using the CRISPR-Cas9 technology do not fall under the European GMO definition. Nevertheless, the European Commission's dilly dallying is extremely damaging for the prospect of

European plant breeders investing in genome-editing technologies, and is a scandalous dereliction of responsibility.

4.7. Labelling and Traceability Regulations

The first GM plant product to come onto the market in the UK was paste made from slow-ripening tomatoes (Section 3.4). This product was approved for food use in the UK by ACNFP in 1995, but the product has not been available since 1999. The paste was sold through two major food retailers, Sainsbury and Safeway, clearly marked with a large label stating that it was made from GM tomatoes (Figure 3.2). The UK food retail industry intended to pursue this policy for all foods derived from GM plants, at least until consumers were familiar with the new technology. However, these plans were thrown into confusion in late 1996 when shipments of that year's harvest of soybean and maize imported from the USA arrived. Both contained at that time approximately 2% GM material with the GM and non-GM all mixed together. With no legislation covering the labelling of GM foods, no agreement with the Americans to supply segregated GM and non-GM produce and insufficient time to identify alternative suppliers (the industry appeared to have been taken by surprise), retailers had a choice: either label everything containing USA soybean and maize as potentially containing GM material or abandon the labelling policy. It chose the latter, a decision that, with hindsight, was undoubtedly a mistake because consumers felt that GM food was being introduced behind their backs.

Labelling controls covering GM foods were finally introduced in Europe in 1997 but not ratified in the UK until 1999. The regulations required that any food containing material from GM crops had to be labelled, with the following exemptions:

- Refined vegetable oils, sugar and other products that do not contain DNA or protein.
- Foods that contain small amounts (below 1%) of GM material as a result of accidental mixing.
- Food sold in restaurants and other catering outlets (the UK opted out of this exemption and required caterers to provide written or verbal information covering GM foods to their customers).
- Animal feed.

The regulations were updated through directives on the regulation of GM food and feed (1829/2003) and the traceability and labelling of GMOs (1830/2003). These regulations, which came into force in April 2004, require the labelling of all food and animal feed containing GMOs or GMO-derived ingredients; so animal feed is no longer exempt. The threshold for labelling is reduced to 0.9% for the accidental presence of GM material, with zero tolerance if the GM material has not been approved for use in Europe. This zero tolerance rule for products that have been approved elsewhere but are not yet approved in Europe is becoming increasingly unworkable as the amount of GM material in imported food and feed increases. The traceability and labelling regulations require a system for tracking GM products through the supply chain from seed company to farm, processors, distributors and retailers. The exemptions for refined products no longer apply, but the regulations do not cover products from GM-fed animals or enzymes produced in GM microorganisms and used widely in the production of cheeses, yoghurts and other foods. A full list of products that have to be labelled is given in Table 4.2.

The regulations covering refined products are difficult to enforce because the products are identical whether they come from a GM or non-GM source and because without the presence of DNA or protein there is no way of confirming a GM origin. It seems odd that a product that is identical to its non-GM counterpart but comes from a GM plant source should be labelled while there is no requirement to label a food that actually contains a modified enzyme from a GM microorganism.

Although the stringent labelling laws undoubtedly made GM crops less attractive for European farmers and the food supply chain, and continue to do so, they may have had the effect of forcing some sectors to face the reality that major commodity-exporting countries have adopted GM crops widely. In 2007, for example, the FSA compiled data on the global production of GM and non-GM soybean, maize, cotton and oilseed rape (Table 4.3), showing the penetration of GM varieties into the market. Although no official number is available, traded commodities are believed to be dominated by GM varieties to an even greater extent, with over 90% of traded soybeans likely to be GM, as is at least 50% of traded cotton and maize. These crops are indispensable for animal feed production, and millions of tonnes of grain and meal are imported into the UK for feed production every year. The implication of this is that it has become

Table 4.2.   GMOs and GM foods that do or do not have to be labelled under EU regulations 1829/2003 and 1830/2003 that came into force in April 2004.

GMO or GM food	Labelling required?
GM plant or seed	Yes
Whole foods derived from a GM plant source	Yes
Refined products such as oil and sugar from a GM plant source	Yes
Foods containing GM microorganisms	Yes
Animal feed containing GM crop products, including derivatives such as oil and gluten	Yes
Fermentation products produced for food or feed use	Yes
Food from animals fed GM animal feed	No
Food made with an enzyme produced in a GM microorganism (some cheeses, for example)	No
Products containing GM enzymes where the enzyme is acting as an additive or performing a technical function	Yes
Flavourings and other food and feed additives produced from GMOs	Yes
Food containing GM ingredients that are sold in catering establishments	Open to interpretation. FSA opinion is yes.

Source: FSA.

extremely difficult to source enough non-GM product to meet demand at an acceptable cost. This same conclusion led to feed for pigs and cattle being made with GM ingredients from the early 2000s, while poultry feed has been made with GM ingredients since 2013.

The fish feed sector of the animal feed market still avoids the use of GM crop products, and the high end of the market also stipulates non-GM feed for cattle, pigs and poultry. This reflects pressure coming back up the chain from retailers, although it is not clear why it is considered important that fish are not fed with feed containing GM crop products while most cattle, pigs and poultry are. The fish feed market is an interesting one because consumers eat fish partly for omega-3 fatty acids. As described in Section 3.8.3.1, the long-chain omega-3 fatty acids that make fish oil so

Table 4.3. Global production of four crops that are indispensable for animal feed production and the amount that is produced from GM varieties.

	Total (tonnes)	GM varieties (tonnes)
Soybean	86 million	58.6 million (68%)
Maize	140 million	35.2 million (25%)
Cotton	34 million	15 million (44%)
Oilseed rape	23 million	5.5 million (24%)
Total for the above 4 crops	283 million	114.3 million (40%)

Source: Figures compiled by the FSA of the UK for 2007.

valuable are made in marine algae and accumulate through the marine food chain; they are not made by fish themselves. Farmed fish, therefore, must have long-chain omega-3 fatty acids in their diet if they are to accumulate them. This is currently provided in marine fish meal, making farmed fish no more sustainable as a source of long-chain omega-3 fatty acids than marine fish, stocks of which are declining. An alternative and sustainable source of long-chain omega-3 fatty acids should become available in the next few years in the form of GM soybean and oilseed rape, and it seems likely that the fish-farming industry will have to use it.

The labelling laws in the USA are quite different from Europe. Products derived from GM plants do not have to be labelled as such. However, products that are significantly different from their conventional counterpart, such as high oleic acid soybean oil (Section 3.8.2), do have to be labelled. In other words, products are labelled according to their properties, not how they were made.

Chapter 5

Issues that Have Arisen in the GM Crop and Food Debate

The controversy over GM crops has spanned the globe and lasted for more than two decades, with no sign of abating. It is arguably one of the great debates of our time, and a testing ground for the ability of science and scientists to overcome irrationalism and scaremongering. In Europe, there is no sign of the battle being won; indeed, so far it has not even been close. As a result, great damage has been done to the European plant biotechnology industry: Syngenta, Monsanto, Aventis (now part of Bayer), DuPont, Unilever and BASF have all closed or downsized plant biotechnology operations in Europe. Researchers have also been put at a great disadvantage. If a scientist in the USA wishes to undertake a field trial of a genetically modified crop, he/she can submit a one-page application online and receive approval the next day. In Europe, scientists have to negotiate a complicated and expensive regulatory process (Section 4.2) and then face a high probability of their field trial being vandalised. Not surprisingly, only a handful of field trials have been undertaken in Europe in recent years.

The influence of European attitudes to GM crops extends well beyond Europe's borders. Europe is nowhere near self-sufficient in food and is a major market for commodity exporters. The European market is so lucrative that the countries that sell into it are extremely wary of anything that might jeopardise their ability to do so. This raises ethical issues: Golden Rice (Section 3.11), for example, which potentially could benefit millions of people in developing countries who suffer from vitamin A deficiency, has been held up for almost two decades by what its inventor, Ingo Potrykus, describes as 'European Attitudes'. It must be said that Europe is not entirely alone: Japan, for example, has been implacably opposed to GM crops, and preserving its market in Japan has been a major reason for China delaying the commercial cultivation of GM rice.

Exactly why European consumers have been so much more fearful of GM crops than other consumers is not a simple question to answer. Opinion polls taken in Asian countries (other than Japan) and in the Americas consistently show that two-thirds or more of the population broadly approve of plant biotechnology and believe that they are likely to benefit personally from its development. Polls in the UK and Europe have consistently shown much less favourable attitudes amongst consumers. It should be noted that poll results depend in part on what question is asked, and the results of the UK's Food Standards Agency surveys on consumer attitudes (Figures 5.1 and 5.2) show less hostility to GM foods amongst UK consumers than might be expected, given the tone of the debate that has taken place in the UK. One possible interpretation of this is that a small minority of people and organisations that are extremely hostile to GM foods have dominated the debate. The data also show that concern about GM foods declined slowly among UK consumers in the first decade of this century, then levelled off (Figure 5.2), and that concern is expressed by only 7% of respondents if they are left to respond spontaneously, compared with 27% who select GM foods as an issue of concern from a list (Figure 5.1). Several other issues concern consumers more than GM foods (Figure 5.1).

Europeans are not naturally 'Technophobic'. Technology continues to develop rapidly on many fronts, from telecommunications to computing and medicine and more, and most developments have been embraced by Europeans. The internet is just one example of a development that has had a far greater impact on society than GM crops, and people in Europe, as elsewhere, have accepted it, taking proportional steps to minimise risks where they see them, for example by using parental controls to prevent their children from accessing inappropriate websites. Most Europeans

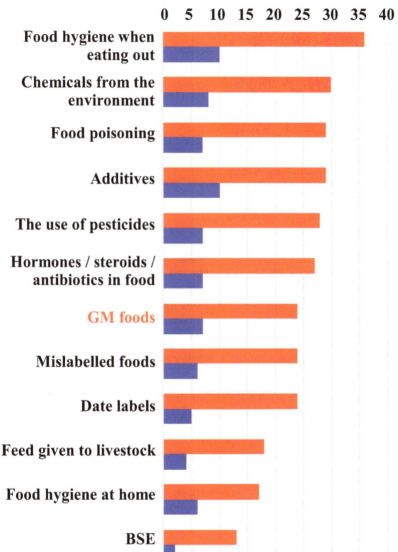

Figure 5.1. Results of the UK's Food Standards Agency survey, 2017, showing main food issues for consumers. Red bars show the proportion of respondents who selected an issue from a list, while blue bars show the proportion of respondents who cited an issue when simply asked what issues concerned them.

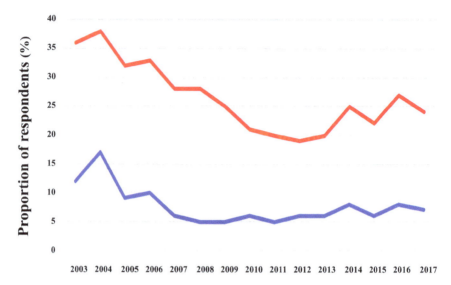

Figure 5.2. **Results of the UK's Food Standards Agency survey of consumer attitudes to GM foods from 2003 to 2017. The red line shows the proportion (%) of respondents who selected GM foods from a list of issues, while the blue line shows the proportion of consumers who cited GM foods when simply asked what food issues concerned them.**

also have a mobile phone, and that market grew astonishingly fast despite concerns being raised by some scientists about their safety when they first became popular. Another development that might have been expected to cause a heated debate was experimentation on human embryos, but while that was considered carefully by the science community and government, there appeared to be relatively little interest amongst the European public. In contrast, experimentation on human embryos has been a hugely controversial issue in the USA, while attempts by pressure groups to ignite a debate on GM crops have largely failed.

That is not to say that pressure groups in the USA have given up. While we in Europe may think of the acceptance of GM crops in the USA as a done deal, the establishment of a market for GM varieties of some crops is still a struggle. Soybean, maize, cotton, oilseed rape and some minor crops have had very successful GM varieties on the market for many years, but potato, for example, has not. Attempts to launch GM insect- and virus-resistant varieties in the 1990s (Sections 3.6.3 and 3.7) failed in the face of a reactionary fast-food industry that was reluctant to adopt any new potato varieties, let alone GM varieties. So, will potatoes such as Innate® and Innate® Generation 2 that are engineered to reduce the potential for acrylamide formation during processing (Section 3.13) be a game-changer? These potato varieties represent the best of modern biotech, carrying multiple traits, including, in the case of Innate® Generation 2, resistance to late blight. However, it is the hugely reduced levels of acrylamide, a carcinogenic, neurotoxic and fertility-affecting processing contaminant, in products produced from the potatoes that may force the fast-food industry and NGOs to reassess the GM issue.

So far that has not happened. The Organic Consumers Association, for example, petitioned McDonalds not to use the potatoes, while Food and Water Watch ran an online petition calling for the potatoes to be boycotted. McDonalds conceded and other fast-food companies and processors fell into line, but Simplot, who developed Innate® and Innate® Generation 2, have focussed on the fresh produce market instead. It is an interesting issue to follow: if fast-food and food processing companies reject GM varieties like Innate® and Innate® Generation 2, how will they deal with their acrylamide problem and the evolving regulations that they will have to comply with? Will they take a similar stance with genome-edited varieties? Was it unwise for NGOs to take a stance against GM varieties with a clear consumer health benefit?

Despite this development, Europe remains the key battleground for the acceptance of GM and, potentially, genome edited crops. Part of the answer as to why that is lies in the reluctance of Europeans to trust their governments or scientific experts on food issues. GM foods were launched in Europe shortly after the epidemic of bovine spongiform encephalopathy (BSE) in the UK cattle herd, which led to one of the biggest food scares in UK history. Rightly or wrongly, consumers felt that they had been given the wrong advice by scientists and government ministers on the safety of beef. Food 'scares' are not unique to the UK and Europe, but the BSE issue did undermine the confidence of European consumers in a way that has not occurred elsewhere. This enabled pressure groups to dominate the debate, and European consumers have been bombarded with inaccurate information, half-truths and wild scare-stories, as well as bizarre and frightening imagery (Figure 5.3). Even if consumers do not believe the more hysterical of these stories, why should they take the risk of buying GM food products?

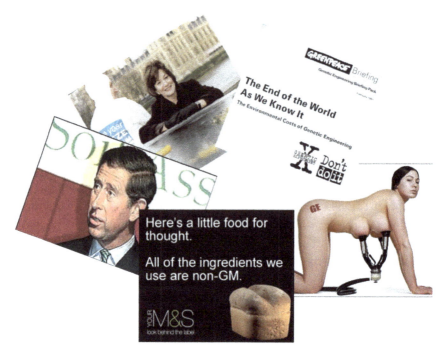

Figure 5.3.   **Images from the GM debate: Clockwise from top left: Joan Ruddock, Member of Parliament and head of the 'Five-year Freeze' campaign, behind a block of ice containing fish heads attached to corn cobs; campaign leaflet from Greenpeace, October 1997; billboard image used by 'Mothers Against Genetic Engineering' in New Zealand in a campaign to prevent a field trial of GM herbicide-tolerant onions; online advert from Marks and Spencer's in 2008 arguably implying that other retailers might be selling GM bread (no GM wheat was or is grown anywhere); Prince Charles at a Soil Association press conference at which he stated that 'Mixing genetic material from species that cannot breed naturally, takes us into areas that should be left to God'.**

Journalists have generally been happy to repeat the proclamations of pressure groups. Some of the wilder headlines from the UK include:

- 'Scientists warn of GM crops link to Meningitis'. *Daily Mail*, 26th April 1999. The article continued, 'The nightmare possibility of GM food experiments producing untreatable killer diseases has been underscored by senior government scientists'. The story was retracted the following day in a short statement on an inside page.

- Are we at risk from mutant make-up? *Express on Sunday*, 21ˢᵗ February 1999.
- Scientists raise the fear of GM foods triggering new allergies. *The Express*, 30ᵗʰ April 1999.
- Lifting the lid on the horror of GM foods. *The Express*, 12ᵗʰ May 1999.
- The GM pollen that can mean a cloud of death for butterflies. *Daily Mail*, 20ᵗʰ May 1999.
- Mutant porkies on the menu. *News of the World*, 23ʳᵈ May 1999.
- GM risk in daily food of millions. *Guardian*, 24ᵗʰ May 1999.
- GM food 'threatens the planet'. *Observer*, 20ᵗʰ June 1999.
- Meat may be tainted by Frankenstein food. *Daily Mail*, 6ᵗʰ July 1999.
- M&S sells genetically modified Frankenpants. *Independent on Sunday*, 18ᵗʰ July 1999.
- Is GM the new thalidomide? *Daily Mail*, 8ᵗʰ October 2003.
- How GM crop trials were rigged. *Independent on Sunday*, 12ᵗʰ October 2003.
- Stop the rush to GM crops. *Independent on Sunday*, 12ᵗʰ October 2003.
- Curb on GM crop trials after insect pollution. *Daily Telegraph*, 14ᵗʰ October, 2003.
- Polluted for generations. *Daily Mail*, 14ᵗʰ October 2003.

Scientists in the UK have been criticised for losing the GM debate, through failing to engage with the public or the media, failing to communicate in terms that non-scientists can understand, or reacting too slowly to issues when they arose. However, it could be argued that there is not a lot that scientists could have done in the face of a campaign that was waged at the level demonstrated by the headlines above. In any case, few scientists are trained to deal with the media or the public, and most find it a difficult and uncomfortable experience. The 'correct' answer to the question 'Are GM foods safe' is the one given in Chapter 4: 'There is an international consensus amongst experts in the field that GM is not inherently more risky than other methods in plant breeding; GM foods are examined under a rigorous assessment system that goes beyond that applied to other foods; the safety of GM food has been considered by the

World Health Organisation, the European Union, the Organisation for Economic Co-operation and Development, as well as many national governments and none has concluded that there is any evidence of adverse effects on health'. However, there is little chance of getting through the first line of this statement before a television interviewer interrupts or the audience starts thinking about something else. The opposing statement that GM foods are unsafe or untested may not be true, but it is much simpler to put across. So should scientists simply answer yes to questions such as 'Are GM foods safe?'. Perhaps more than two decades of safe use of GM crops and consumption of GM foods by hundreds of millions of animal and human consumers without the report of so much as a bellyache justifies a 'Yes' answer, but it is a difficult ethical issue for scientists.

It is also a difficult issue for scientists because they believe that if they win the scientific argument the problem will go away. In fact, the scientific debate was held and won decades ago, but this has had no impact whatsoever on the public debate, and very little on politicians, who are unwilling to pay the political cost of relaxing the regulations on GM crops.

So, what are the issues and concerns that have been raised during the GM crop debate? The rest of this chapter describes the major issues, and attempts to address some of them. As someone who has been involved in the debate since the start, it now reads like something of an historic document, and there is no doubt that the debate is of interest for students of sociological and political history as well as crop and agricultural science.

5.1. Are GM Foods Safe?

As discussed above, scientists are always reluctant to describe anything as completely safe. However, if we accept that new crop varieties produced by non-GM methods are safe enough, it is reasonable to judge the safety of GM crops in comparison. The major arable and horticultural crops grown and consumed in the world today have been developed over centuries or even millennia. Consequently, they are assumed by the consumer to be safe and wholesome. However, most, if not all, of these crops contain compounds that are potentially toxic or allergenic. In most cases, these compounds have probably evolved to provide protection against animal grazers or pathogenic microorganisms and it is therefore not surprising that they are also toxic to humans. They include glycoalkaloids in potatoes, cyanogenic glycosides in linseed, glucosinolates in *Brassica* oilseeds and proteinase inhibitors in soybean and other legume seeds. Some of these have been discussed previously in this book.

It is inconceivable that these or many other generally accepted foods would be approved for food use had the toxins been introduced by genetic modification. Nevertheless, with some exceptions, the introduction of new types and varieties of food crops produced by conventional breeding requires no specific testing for the presence of allergens or toxins, even if genes have been introduced from exotic varieties or related wild species by wide crossing.

The products of genes introduced by genetic modification are readily identified in a GM plant. They may also be isolated in a pure form, either from the species of origin, from the GM plant or after expression in a microorganism. The pure protein can then be tested in detail and its presence in processed foods monitored. In contrast, it is virtually impossible to predict or characterise all of the changes in food composition that may result from conventional plant breeding. As for unpredictable effects, as described in Chapter 4, environmental factors have been shown to have more effect on the composition of wheat grain, for example, than genetic modification, while gene expression differs much more between different genotypes produced by conventional breeding than between different GM genotypes.

On top of all this, as described in Chapter 4, GM crops and foods are examined under a rigorous assessment system. Consequently, it would not be unreasonable to argue that GM foods may be safer than food derived from new, non-GM varieties.

5.2. Will Genetic Modification Produce New Food Allergens?

It has been suggested that consumption of GM foods could lead to increases in allergenicity. In fact, there has already been a widely reported case of increased allergenicity in a GM plant line, but the problem was detected and the product never reached the market. In the early 1990s, a methionine-rich 2S albumin storage protein gene from Brazil nut was expressed in soybean in order to increase the content of methionine, an amino acid, in soybeans used for animal feed. The protein was subsequently shown to be an allergen, as are a number of related 2S albumins from other species, and the programme was discontinued.

As with the more general question of safety, it could be argued that genetic modification is no more or less likely to lead to an increase in food allergens than other methods in plant breeding. Nevertheless, the industry and regulators are clear that every precaution should be taken to ensure that no new allergens are introduced into the food chain through genetic modification.

In fact, the biotechnology industry is bullish about its ability to predict whether or not a GM food is likely to be more allergenic than its traditional counterpart and to detect new allergens before a product reaches the market if they get their predictions wrong. A typical procedure described by Astwood and co-workers (*Food Allergy*, 2nd Edition, Blackwell Science, 1996) combining molecular analyses of the novel gene and protein, immunoassays and skin prick tests is shown in Figure 5.4.

Figure 5.4. Decision tree showing the assessment and testing of possible allergenicity in GM foods, as described by Astwood and co-workers (*Food Allergy*, 2nd Edition, Blackwell Science, 1996).

5.3. Is it Ethical to Transfer Genes Between Different Species? Are Cisgenics More Acceptable than Transgenics?

This question of the ethics of transferring genes between different species has been raised famously by Prince Charles. He has been quoted as saying that 'Mixing genetic material from species that cannot breed naturally, takes us into areas that should be left to God'. Of course, plant breeders were mixing genetic material from sexually incompatible species long before genetic modification was developed (Section 1.8.1). Nevertheless, this statement encapsulates the views of a significant number of people. It is a pity that the point could not have been made without the use of overblown language, such as the term 'Frankenstein Foods'.

The philosophical counter-argument to this view is that all life on Earth has a common ancestor if you go back far enough, and that the species alive today are part of an evolutionary continuum, not separate entities. So genes encoding proteins involved in cellular control mechanisms that evolved long ago, such as the control of cell division, for example, are closely related and instantly recognisable to those that work on them, whether they come from yeast, invertebrates, plants or humans. So how can it be fundamentally wrong to move genes between species?

The pragmatic counter-argument is that it is not where a gene comes from or how it is brought into a plant breeding programme that is important but what it does. A gene producing a poison that is produced inadvertently by radiation mutagenesis or crossed into a crop plant from a wild relative is clearly much more dangerous than a gene producing a benign protein that is introduced into a crop plant by genetic modification. Similarly, genes imparting herbicide tolerance can 'flow' into weed plants by cross-pollination regardless of whether they have been engineered or crossed into a crop plant, or produced by radiation mutagenesis.

This argument, which has a sound scientific basis, would lead to the conclusion that GM crop varieties should be assessed and regulated no differently to new crop varieties produced by other means, but that is not the case. For example, as described in Section 3.5, GM oilseed rape varieties that are tolerant to one or other of two broad-range herbicides, glyphosate and gluphosinate, are grown widely in North America. There is another broad-range herbicide-tolerance trait available in commercial oilseed rape varieties in Canada, the herbicide in this case being imidazolinone. This trait was developed by Pioneer Hi-Bred, now part of DuPont, but it was produced by mutagenesis, not genetic modification. Oil or meal produced from this variety can be imported into Europe without having to pass through the regulatory system that the GM varieties have had to go through, and without any of the safety assessments. Furthermore, although imidazolinone is not used in Europe, in theory there is nothing to stop a plant breeder crossing this trait into a European variety and marketing it in Europe, again without having to go through the regulatory system that so hinders the development of GM varieties. It is difficult, therefore, to argue that the system is science-based.

Part of the reaction against moving genes between species is an emotional one. The use of an 'antifreeze' gene from a fish to engineer frost resistance into tomatoes, for example, caused a much greater reaction than the use of a bacterial gene to engineer insect resistance into maize, despite the fact that the tomatoes were not a commercial product whereas the maize was. So far, biotechnology companies have avoided using animal genes in GM crops for food use but are using them to make pharmaceutical products in GM plants. Clearly, the industry believes that consumers have different attitudes when they are buying and using pharmaceuticals to when they are buying and consuming food.

If the ethics of transferring genes between species are contentious, what should we make of GM plants that have been engineered with genes entirely from the same species? Such plants have been called *cis*genic, rather than *trans*genic, and a paper published in 2006 by Henk J. Schouten, Frans A. Krens and Evert Jacobsen of Plant Research International, Wageningen University and Research Centre in the Netherlands, argued that cisgenic crops should not come under GM regulations. This argument was rejected by Europe's regulators, but Simplot, who developed Innate® and Innate® Generation 2 (Section 3.13), use the fact that these varieties are cisgenic, in other words all of the genetic modifications were made using potato DNA, in their promotional material. The question then arises as to how close a relative to a crop species must a source species be to qualify as a cisgene rather than a transgene? The late blight resistance gene, *Rpi-vnt1.1*, that is present in Innate® Generation 2, for example, originated from a wild potato, *Solanum venturii* (Section 3.12): definitely a potato species, but not the same as cultivated potato, *Solanum tuberosum*.

5.4. Animal Studies

The World Health Organization is sceptical about the usefulness of animal feeding studies in the safety assessment of GM plants and foods. This scepticism arose in part from experience gained in the testing of irradiated food in the early 1990s. Animal studies are undoubtedly useful in the toxicological safety assessment of individual compounds such as pesticides, pharmaceuticals, industrial chemicals and food additives. It is relatively simple to feed such compounds to animals at doses sometimes far higher than humans would be exposed to in their diet and to identify any potential adverse effects on health. Foods, on the other hand, are complex mixtures of compounds. Animals cannot be persuaded to eat orders of magnitude more of them than humans can, and feeding only one type of food to an animal usually reduces the nutritional value of the diet, causing adverse effects that are not related directly to the material itself. For these reasons, relating any effects on the welfare of an animal to a particular genetic modification can be extremely difficult.

Nevertheless, the World Health Organisation recommends the use of animal studies if other information available on the GM plant or food is inadequate, particularly if the novel protein in the GM food has not been present in the food chain before. It also supports the use of animal testing of proteins produced by novel genes before the gene is used in biotechnology (in other words, for the gene to be engineered into a microorganism so that large amounts of its protein product can be made and purified; the individual protein can then be used in toxicology studies).

The fact that GM foods did not have to be tested on animals was used by pressure groups in the late 1990s and early 2000s to undermine consumer confidence in GM food safety. Partly for this reason, all applications made in Europe for consent to market foods produced from GM crops now include the results of animal feeding experiments, despite the ethical question of whether it is right to undertake studies on animals if the results are unlikely to be meaningful.

5.5. GM Crops 'Do Not Work'

A long-running strategy of anti-GM campaigners is to claim that GM crops are flawed in some way. In other words, there is a problem inherent in the technology that means that no GM crop will deliver what it promises. GM crops were linked by activists with crop failures in India and Indonesia in the 1990s, for example, despite the fact that neither of these countries was using GM crops at that time, and with farm failures in the USA, where farmers continue to use GM crops in increasing numbers.

In its report, 'Seeds of Doubt', the Soil Association, a UK organic farming group, claimed that the use of GM crops in the USA had reduced profitability, reduced yields and raised costs through increased herbicide use. Herbicide-tolerant soybean was cited as an example of a crop that had failed to meet expectations of higher yields. In contrast, a much more extensive report from the National Centre for Food and Agricultural Policy cited herbicide-tolerant soybeans as the most successful GM crop in improving farm profitability. Increased profits were being gained through reduced costs, not improved yield. Indeed, improved yield was never a target for herbicide-tolerant crops in a developed country where farmers have access to a variety of weed control measures. So why do the two reports differ? Perhaps because the Soil Association report, as it states itself, did not cover positive experiences.

Not all new crop varieties, GM or otherwise, are successful, and some GM crop failures are described in Chapter 3. Indeed, one of the first GM varieties on the market, the 'Flavr Savr' tomato (Section 3.4), was a commercial failure and was soon withdrawn. However, some GM crop varieties have been extremely successful. In 1996, when GM herbicide-tolerant soybean was first grown widely, few people even in the biotechnology industry would have predicted that by 2009 more than three-quarters of the world's soybean crop would be GM. It is not the purpose of this book to 'push' the use of GM crop varieties in favour of others. However, it is appropriate to point out that the major GM crop varieties have now been available around the world for two decades, long enough for farmers to make a judgement. The continued popularity of these varieties indicates that farmers find some benefit in using them. Independent data on crop yields and production are not difficult to find, and the data on cotton in India shown graphically in Figure 3.5, for example, tell their own story.

5.6. Did Tryptophan Produced by Genetic Modification Kill People?

This issue was often brought up when GM crops first came onto the market, although it concerns GM bacteria, not plants. Tryptophan is an amino acid and is essential to all living things in order for them to make proteins. Plants, fungi and bacteria make their own, but animals, including humans, do not, and need to acquire tryptophan in their diet. Experiments with rats fed tryptophan-deficient diets suggest that tryptophan deficiency delays growth, development and maturation of the central nervous system. As well as its role as an amino acid required for protein synthesis, tryptophan is a precursor for a neurotransmitter, serotonin. There is no evidence that tryptophan deficiency is a problem in developed countries, since it is plentiful in meat and dairy products and a variety of plant sources, but health supplement manufacturers claim an extensive range of benefits associated with taking tryptophan as a supplement.

Many pharmaceutical companies around the world therefore manufacture tryptophan, usually from selected strains of bacteria. One of these companies, Showa Denko in Japan, used a GM strain of *Bacillus amyloliquefaciens* to produce the raw product in the late 1980s. In 1989, the company changed the strain of bacterium that it was using and reduced the number of purification processes that it used to clean up the raw product. Subsequently, certainly hundreds, probably thousands, of people in the USA who bought the product under various brand names became ill with *Eosinophilia myalgia syndrome* (EMS), with symptoms of elevated counts of blood eosinophils (a type of white blood cell) and myalgia (severe muscle pain). Tragically, 38 people died. Tryptophan was withdrawn as an over-the-counter product in the USA and the Food and Drug Administration (FDA) has never allowed its return, although it can be obtained on prescription.

An investigation into the incident found that several contaminants that should have been removed in the purification process were present in the finished product. Why they were still present is not clear since Showa Denko did not release the results of its own internal inquiry. It is also still not certain which contaminant was responsible for the illnesses and deaths. To my knowledge, no manufacturer is currently using GM bacteria to make tryptophan, but problems with contaminants and the resulting condition of EMS still occur sporadically, which suggests that the problem is one of old-fashioned chemical engineering, not genetic modification.

The irony of this story for plant biotechnologists is that although it concerns GM bacteria, it was used as an argument against the genetic modification of plants. Meanwhile, the use of GM bacteria to produce medicines, vaccines, food additives and supplements attracts little attention.

5.7. The Monarch Butterfly

In 1999, a study conducted by John Losey and his team at Cornell University found that caterpillars of the monarch butterfly that were forced under laboratory conditions to eat large quantities of pollen from GM insect-resistant maize suffered higher mortality levels than caterpillars that were not fed the pollen. The study was published in the journal Nature and caused an international furore. Indeed, because of the iconic status of the monarch butterfly in the USA, there were fears that the study might cause the sort of public backlash against GM crops there that had occurred in Europe. As it turned out, although the issue was more grist to the mill for the media and anti-GM activists in Europe, Americans remained broadly supportive of plant biotechnology.

The *Cry1A*, or Bt gene (Section 3.6) that is engineered into maize to make it insect-resistant is toxic to caterpillars. The gene is designed to be active everywhere in the plant, including the pollen, so it is no surprise that monarch butterfly caterpillars that ate the pollen did not thrive. However, this was a laboratory experiment. In the wild, monarch butterfly larvae eat milkweed, not maize pollen. In the experiment, the pollen was spooned onto milkweed leaves so that the larvae had no choice but to eat it. Even as the first report was published, experts were extremely sceptical that maize pollen would ever accumulate in such amounts on milkweed in the wild. Field-based studies published subsequently bore this out.

Similar laboratory-based experiments have shown that the survival rate of predator species such as lacewings and ladybirds can be reduced in laboratory experiments if they are fed exclusively on prey species that are feeding on GM insect-resistant plants. None of these results have been replicated in the field. It should be remembered that spraying caterpillars, lacewings, ladybirds and other insects with pesticide, which is the practice for non-GM maize, kills them all outright. The use of GM insect-resistant plants in the field can, therefore, lead to an increase in beneficial insects. With GM maize, for example, the farmer does not have to use an early spray against the corn borer. The farmer may need to spray against other pests but, if this can be avoided, benign predatory insects can thrive. If this happens, the predatory insects prevent a late infestation of red spider mites, reducing pesticide use even further.

As soon as the initial report was published in 1999 and the controversy erupted, Losey was at pains to point out that the study was a preliminary one and that field studies would be required to find out the real impact of GM crops on butterfly populations. This raised eyebrows amongst the science community because Nature does not normally publish preliminary reports. Indeed, many scientists felt that the study would not have been considered worthy of publication without the GM crop tag and that publication was rushed through in order to generate headlines and publicity.

5.8. The Pusztai Affair

The other incident that brought the GM issue to the fore in the late 1990s involved Dr Árpád Pusztai, who was working at the Rowett Institute in Scotland, UK. Dr Pusztai had produced GM potatoes that were engineered to produce a type of lectin. Lectins are proteins that bind carbohydrates and have a range of functions in nature. Many plants produce them as natural insecticides, but most lectins are unsuitable for use in plant biotechnology because they are poisonous to animals, including humans. The lectin gene that Pusztai used came from snowdrop. When the GM potatoes were fed to rats, the rats suffered from lectin poisoning, which is, perhaps, not very surprising. However, Pusztai claimed to have fed another group of rats with ordinary potatoes that had been spiked with the same lectin.

These rats, he said, suffered from lectin poisoning but in his view they did not suffer as badly as those fed with the GM potatoes. Pusztai therefore claimed that the process of genetic modification had somehow made the potatoes more poisonous than they would have been otherwise.

The normal method of dissemination of scientific results is through the scientific literature. Every article is reviewed by at least two independent experts in the field (peer review), the aim being to make sure that experiments have been conducted properly and that the deductions made by the authors of the article are justified by their data. This system is not infallible but no one has come up with a better one. However, on 10th August 1998, before the study had been submitted for peer review, a documentary in the *World in Action* series was broadcast on national television in the UK in which Dr Pusztai claimed that the effect of feeding GM potatoes to rats 'was slight growth retardation and an effect on the immune system'. On the same day, a Rowett Research Institute press release made the same claims and called for further research.

There was huge, international media interest in the claims, and this was heightened even further when, 2 days later, the Rowett Institute published another press release announcing that Dr Pusztai had been suspended and would be retiring. It turned out that some of the experiments that had been referred to in the *World in Action* programme had not been carried out, but the announcement of Dr Pusztai's suspension and retirement led to speculation in the media that his evidence of potential dangers of GM foods was being suppressed.

The affair became such an important public issue that the experiments and results were reviewed by a group of scientists appointed by the Royal Society, who concluded that the work was 'flawed in many aspects of design, execution and analysis' and that 'no conclusions should be drawn from it'. Nevertheless, a medical journal, *The Lancet*, published the study in 1999, giving it a degree of credibility that it did not deserve.

The incident was considered by the UK's Parliamentary Select Committee on Science and Technology in its inquiry into genetic modification. The report of the inquiry is available at:

www.parliament.the-stationery-office.co.uk/pa/cm199899/cmselect/
cmsctech/286/28605.htm

Dr Pusztai admitted to the inquiry (paragraph 26 of the report) that 'no differences between parent and GM potatoes could be found', directly contradicting his statement in the *World in Action* programme. Nevertheless, as the report goes on to state, the press continued to give credibility to Dr Pusztai's claim, despite it being contradicted by his own evidence, and still does. Pusztai became a hero of anti-GM campaigners, to whom he acted as a consultant long after his retirement.

5.9. Alarm Caused by Contradictory Results of Biosafety Studies

The Pusztai study is not the only one to have produced results contradicting the scientific consensus on GM crop and food safety and to have caused great controversy after being publicised without scientific peer review. Herbicide-tolerant soybeans, for example, were approved for food use in the 1990s and have been widely eaten and studied since then, with no alarm being raised. However, in October 2005, Dr Irina Ermakova of the Institute of Higher Nervous Activity and Neurophysiology of the Russian Academy of Sciences reported that she had conducted a study that showed much higher death rates and lower body weight in young rats whose mothers were fed a diet of herbicide-tolerant GM soybeans compared with rats fed non-GM soybeans. The results were reported without being scientifically peer-reviewed.

Toxicology experts, including those on the relevant UK expert advisory committees and the European Food Safety Authority, complained that they were unable to perform a serious review of the study because insufficient information was available on how the experiments had been conducted. The UK's Advisory Committee on Novel Foods and Processes (ACNFP) stated that they considered the results sketchy and inadequately supported, with no information on the composition of the rats' diets. They also pointed out that the study's findings contradicted peer-reviewed scientific studies that had not shown any negative health effects. For example, Denise Brake and Donald Evenson at South Dakota State University had conducted similar feeding studies on mice and published their results in a peer-reviewed scientific journal, *Food and Chemical Toxicology*, in

2004. They had found no negative effect of the GM soybeans. That did not stop the Russian newspaper Pravda predicting that life expectancy would decrease as a result of the widespread consumption of GM soybeans, or the UK's Daily Mail warning of risks posed to unborn babies.

GM maize has also come under attack, again despite being approved and used in food and feed since the mid-1990s. In 2007, a French scientist, Gilles-Eric Séralini and colleagues at the University of Caen in France, re-analysed and questioned the interpretation of results of animal feeding experiments with GM insect-resistant maize variety MON863. MON863 carries a Bt gene imparting resistance to the corn rootworm. Séralini had founded the Committee of Research and Independent Information on Genetic Engineering (CRIIGEN), which opposes GM crops and food, in 1999, and this project was funded by Greenpeace. He claimed that the data on rats fed MON863 grain showed effects on body weight, triglyceride levels in females, and urine composition in males, as well as damage to the liver, kidney, adrenal glands, heart and haematopoietic system. The French government used this study as an excuse to stop the growing of GM maize in France, in defiance of the European Commission, even though the variety being cultivated was MON810, not MON863. GM maize cultivation had been gaining popularity with farmers in France in 2006 and 2007 but had become increasingly controversial. The French government has maintained this ban ever since, despite the fact that EFSA reviewed Séralini's study and found that all the blood chemistry and organ weight values for the GM maize-fed rats were actually within the normal range, and that incorrect statistical methods had been used. Séralini's study was also criticised by one of France's own expert committees, the Commission du Génie Biomoléculaire (CGB).

The following year, 2008, the Austrian government released the results of a study, without peer review, from Jürgen Zentek at the University of Vienna, that claimed to show fertility problems in mice fed a GM maize hybrid produced by crossing MON810 with the glyphosate-tolerant line, NK603 (the hybrid therefore carrying stacked insect resistance and herbicide-tolerance traits). The publication of the report caused considerable controversy and was used by Austria and other countries to justify their bans on GM crops and foods. However, the Austrian government announced in October 2009 that they were withdrawing the study, saying that Zentek and his co-workers had failed to deliver a satisfactory report on it, especially with respect to the statistical analysis of the data, and that the government no longer expected to receive such a report. It is deeply regrettable, of course, that the Austrian government did not check the facts or wait for peer review before going public.

In 2009, Séralini was back again, publishing another study, again based on a re-analysis of toxicity data, this time on GM maize lines NK603, MON810 and MON863. Séralini claimed that the data contained evidence of liver, kidney and heart damage in rats fed the GM maize. EFSA again reviewed the study and again concluded that Séralini's claims were not supported by the data and that the statistical analysis was flawed. The French Haut Conseil des Biotechnologies (High Council of Biotechnologies; HCB) also reviewed the study, and concluded that it presented 'no admissible scientific element likely to ascribe any haematological, hepatic or renal toxicity to the three re-analysed GMOs.' The HCB also questioned Séralini's independence. Nevertheless, Séralini continued, and in 2011 published another review of animal feeding studies, concluding that GM food had sex- and dose-dependent liver and kidney effects.

Then in 2012, Séralini published in the journal *Food and Chemical Toxicology* the results of a 2-year study of rats fed GM maize and glyphosate. He claimed that the rats fed the GM maize and glyphosate were more likely to develop tumours than rats fed a control diet. The study received widespread criticism from scientists and regulatory authorities, in particular for the breed of rat, Sprague Dawley, that was used. Sprague Dawley rats have been bred for use in medical research, and develop tumours at a high rate over their lifetime regardless of the diet they are fed. In fact, the tumours that the rats in the study developed were typical of those normally seen in that breed of rats, and described in the information provided by suppliers. The French Society of Toxicologic Pathology therefore stated that the inclusion in Séralini's article of images of tumours from rats fed the GM maize and glyphosate diet, without also showing tumours from rats fed a normal diet, was misleading. *Food and Chemical Toxicology* retracted the paper in November 2013, describing it as inconclusive and unreliable. However, in June 2014 a revised version of the article was re-published by *Environmental Sciences Europe*.

Meanwhile, in April 2010, EFSA reported on its re-evaluation of the GM maize varieties that were approved for food and feed use in Europe, confirming that it considered there to be no scientifically based cause for concern. However, these incidents and the furore that surrounded them only go to show once again that the views of Europe's own expert scientific committees, and those of individual Member States, are largely ignored by the European public and, if it suits them, European politicians and regulators.

5.10. 'Superweeds'

The term 'superweed' is an example of the emotive hyperbole that has dominated the debate on GM crops in Europe. It arises from the notion that a gene that imparts herbicide tolerance in a GM crop could 'flow' into related wild species through cross-pollination, resulting in hybrids that could become uncontrollable weeds.

The potential environmental impact of cross-pollination of GM crops with wild species has to be assessed case by case, taking into account the species and genes involved. Maize and potato, for example, do not cross with any wild species in the UK (although forced crosses can be made between potato and black nightshade in the laboratory) and wheat does not cross with any native plant species to produce fertile hybrids. Wheat is also almost entirely self-pollinating. Oilseed rape, on the other hand, will cross with other cultivated and wild *Brassicas*, including Chinese cabbage, Brussels sprouts, Indian mustard, hoary mustard, wild radish and charlock. The extent of such crossings in agricultural systems is the subject of continuing research, but it does not necessarily mean that GM oilseed rape represents a problem, just that the potential risk represented by GM oilseed rape is different to that of wheat. Studies carried out in Canada, where there are millions of hectares of GM herbicide-tolerant oilseed rape, have not found any hybrids between the GM crop and wild relatives. Even if a hybrid between a GM crop plant and a wild relative did arise, it would only prosper if the gene involved could confer a competitive advantage, so herbicide-tolerance genes would be unlikely to persist in the wild.

Another way in which herbicide tolerance could arise in weeds is if the heavy selective pressure imposed through the widespread use of one herbicide could lead to weed species evolving tolerance to that herbicide independently. This has already been discussed in Section 3.5.4, and the evidence that glyphosate-tolerant weeds were appearing in the USA led Monsanto to develop its 'Roundup Ready Xtend Crop System' based on GM varieties carrying stacked glyphosate and dicamba tolerance traits. Stacking herbicide traits in this way enables farmers to deal with glyphosate-tolerant weeds and makes it less likely that weeds will be able to develop resistance to the herbicide regime.

If the problem of weed populations acquiring herbicide tolerance were to become too serious, the herbicide and the GM crop that went with it would become useless. Farmers would have to turn to other varieties and different herbicides. Farmers have been aware of these problems since herbicides became used widely, long before genetic modification was developed. Farmers will also have to be careful that GM crops with different herbicide-tolerance traits do not cross to produce multi-tolerant hybrids.

5.11. Insect Resistance to Bt Crops

Just as the potential for weeds to develop resistance to control measures is never far from the minds of farmers and those involved in agrochemical production, so is the potential for insect pests to develop resistance to pesticides. This predates genetic modification, but the introduction of GM plants containing their own insecticide brought a different aspect to this old problem. In the USA, the responsibility for monitoring and controlling GM plants that had been engineered to be insect-resistant through the introduction of a *Cry* gene (Section 3.6) was given to the Environmental Protection Agency (EPA). The EPA was already responsible for controlling the use of the Bt pesticide and argued that since both the GM plants and the pesticide used the same insecticidal protein the two had to be controlled by the same agency. This makes sense in some ways, but other

GM crops are controlled by the Animal and Plant Health Information Service (APHIS) within the US Department of Agriculture, and this splitting of responsibility for GM crops may have contributed to the StarLink fiasco (Section 5.17).

The Bt pesticide is an important one for organic and salad farmers in the USA and elsewhere. Organic farmers because the Bt pesticide is allowed by most organic accreditors and salad farmers because Bt is considered safe enough to allow its use right up to the point of harvest. However, the emergence of resistance to the pesticide was considered to be a significant risk, and this had been a long-term concern for the EPA. In response to the introduction of Bt GM crops in 1996, the EPA insisted that farmers using GM crops containing the *Cry1A* gene would have to plant a proportion of non-GM crop as well. This would provide a refuge in which insects that had developed resistance to the effects of the Bt protein would not be at a selective advantage (in fact they would be at a selective disadvantage). According to the EPA's predictions of how the insect population would respond, the proportion of non-GM crop would have to vary according to what other insect-resistant GM crops were being grown in a particular area.

If the *Cry1A* gene were to 'flow' into weed plants through cross-pollination, the models on which the EPA bases its predictions of how insect populations are likely to respond would break down. For this reason, the EPA proposed banning GM insect-resistant cotton in areas of the USA where cotton has wild relatives. As is always likely in the USA, this proposed interference from a Federal Government Agency in Washington in what varieties farmers could or could not grow went down very badly in the Mid-West. In response to criticism from farmers' representatives, the EPA published a report that stated that if its recommendations were not followed there would be widespread resistance to Bt in insect populations within 3–5 years. Both the insect-resistant GM crops and Bt pesticides would then be useless. The EPA got its way and the refuge policy was introduced. So far, more than two decades later, it appears to have been very successful. However, activists in Europe lifted the worst-case scenario prediction out of the EPA's report and used it as one of the justifications of their opposition to GM crops, predicting the emergence of 'superbugs'.

5.12. Segregation of GM and Non-GM Crops: Co-existence of GM and Organic Farming

Unless GM food is accepted universally, which seems unlikely in the foreseeable future in Europe, it is important that alternatives remain available to allow consumers to exercise choice. For imported food-stuffs, European suppliers will either have to buy produce from countries where GM crops are not grown, which has become difficult for some crops, such as soybean, or pay farmers overseas to grow non-GM varieties. This means paying a guaranteed price to a farmer to use old-fashioned varieties and high chemical inputs.

If GM crops are to be grown widely in Europe, one of the key issues that will arise is that of segregation of GM and non-GM crops and food. Segregation could break down through accidental mixing of GM and non-GM seed for planting, cross-pollination between GM and non-GM crops (not a problem for in-breeding species such as wheat), or mixing of the product between the farm gate and the consumer. Some inadvertent mixing is almost inevitable, and the production of totally GM-free food is therefore likely to be expensive and difficult.

The organic farming industry in Europe and the USA has attempted to corner the GM-free market. Indeed, presenting itself as GM-free has been an extremely successful marketing strategy even in countries where other farmers are not using GM crops either. As part of this strategy, the organic industry persuaded authorities in the USA and Europe to write GM-free into the official definition of organic produce. This has led to the bizarre situation where an organic farmer can spray Bt pesticide onto a crop and still sell it as organic, but cannot grow a crop that has been engineered to produce the Bt protein itself.

Clearly, if organic and GM farming are to co-exist there will have to be tolerance on both sides. Organic farmers will have to accept that there might be some cross-pollination of their crops from GM varieties planted in adjacent fields. On the other hand, farmers who have paid for expensive GM seed in order to produce a high-value product will have to accept that some of the seed produced by the crop will result from cross-pollination from old-fashioned varieties grown by the organic farmer next door. Studies in Australia have suggested that a tolerance level of 1% (i.e. a crop with 1% GM content could still qualify as organic) would be sustainable even for out-breeding crops such as oilseed rape. It should also be remembered that there is considerable experience in the UK and elsewhere in Europe of segregating different crop varieties. A good example is the segregation of oilseed rape varieties suitable for industrial uses from those bred to be used for food (Section 3.8), something that has been done successfully for several decades.

Organic farmers represent a tiny minority of the total, selling into a niche market with relatively affluent consumers. Crops with products for which there is no mark-up associated with an organic label are simply not grown organically. Almost no organic sugar beet or oilseed rape is grown in the UK, for example, because there is no market for it. While the interests of organic farmers are important, therefore, self-imposed rules on the incompatibility of GM crops with organic farming should not be allowed to affect the availability of GM varieties for farmers producing food for the wider public.

5.13. Antibiotic Resistance Marker Genes

Another topic that has generated a great deal of debate is the use of antibiotic resistance genes as selectable markers. The use of marker genes to select cells that have been modified with genes of interest is discussed in Section 2.8, and antibiotic resistance genes have been extremely valuable in the development of GM technology. Many scientific bodies around the world, including the World Health Organization and regulatory committees set up by the European Union and several national governments, have considered the safety of antibiotic resistance genes in food and have concluded that those that are being used do not represent a health threat. The British Medical Association, however, have expressed reservations and the UK's ACNFP has called for the development of alternative marker systems. In the late 1990s, the biotechnology industry came to the conclusion that while antibiotic resistance marker genes were safe they were a public relations liability and, while they have continued to be used in basic research, I am not aware of such genes being used in any commercial GM varieties since then. At least, that was the case up to 2017, when Arctic® Apples (Section 3.14) began to be marketed. These apples contain a kanamycin resistance marker gene, along with a gene to prevent oxidative browning.

The supposed risk associated with the use of antibiotic resistance marker genes is that they will somehow find their way into gut or soil bacteria and from there into disease-causing bacteria. In fact, the antibiotic resistance marker genes used in plant biotechnology confer resistance to antibiotics such as kanamycin and neomycin that are not used at all in oral medicines. Furthermore, a gene such as the kanamycin resistance gene in Arctic® Apples is designed to work in the plant and would not be active in bacteria.

Antibiotic resistance marker genes are also used in the genetic modification of bacteria, including bacteria used for the cloning, manipulation and bulking up of genes to be used in plant transformation. In this case, such a gene is designed to be active in bacteria. Where whole plasmids are used to transform plant cells by particle bombardment (Section 2.5), such a gene may be incorporated into the plant genome along with the gene of interest. The antibiotic resistance marker genes used for the genetic modification of bacteria include one that imparts resistance to ampicillin, which is used in medicine. This gene was reported to be present in an early variety of GM insect-resistant maize but is not present in any of the varieties used today.

So, is there any risk associated with the presence of such a gene in a GM plant? Firstly, the gene is integrated into the genome of the GM plant and the risk of horizontal gene transfer from plant genomes to soil or gut bacteria under natural conditions is extremely low, if not non-existent. Furthermore, where did these genes originate from in the first place? The ampicillin resistance gene came from a human gut bacterium, *E. coli*. In fact, even a cursory examination of the information available shows that many bacterial species naturally contain an ampicillin resistance gene, including *Kluyvera ascorbata*, *Pyrococcus furiosus*, *Proteus mirabilis*, *Bacillus subtilis*, *Klebsiella pneumoniae* strain H18, *K. pneumoniae* strain G122, *Pseudomonas aeruginosa*, *P. furiosus*, *Staphylococcus aureus*, *Synechocystis* sp. PCC 6803, *Sinorhizobium meliloti*, *Yersinia pestis* strain CO92, *Mycobacterium leprae*, *Deinococcus radiodurans*, *Vibrio cholerae* and many soil bacteria. In other words, the 'scare' over antibiotic resistance genes used in plant biotechnology is that they could somehow find their way into bacterial populations that already have them.

Antibiotic resistance marker genes are only maintained in bacterial populations if they impart a selective advantage. Antibiotic resistant strains of pathogenic bacteria do represent a health threat, but they arise naturally and thrive because of the sloppy management of antibiotics in human and animal medicine, not because of the use of antibiotic resistance marker genes in crop biotechnology.

5.14. Patenting

Patenting law was devised for two reasons: firstly to allow inventors to benefit from their inventions, and secondly to bring inventions into the public domain so that they would not 'die' with their inventors. The invention must be described in sufficient detail to allow others to repeat it and must satisfy three criteria for the patent to be approved: it must be novel, it must be useful and it must be 'non-obvious' even to an expert in the field.

All of this seems quite laudable, and the system has generally achieved its objectives. Patents protect inventions for up to 20 years, enabling inventors to prevent competitors from copying their invention or to license use of the technology at a price. Any new gadget is likely to be covered by a plethora of patents and licensing agreements.

Traditionally, patents were not awarded to new plant varieties or to strategies devised by plant breeders to improve crops. Indeed, until relatively recently, plant breeders had almost no commercial protection at all for their new varieties. This changed in 1961 with the advent of Plant Breeders' Rights, a form of intellectual property designed specifically to protect new varieties of plants. Plant breeders' rights were drawn up at The International Convention for the Protection of New Varieties of Plants (the UPOV Convention) in Geneva in 1961 and became law in the UK in 1964. They were revised and strengthened in 1991, although the new regulations did not come into force in the UK until 1997.

Plant breeders' rights enable the holder of the rights to prevent anyone from producing, reproducing, offering for sale or other marketing, exporting, importing, conditioning for propagation or stocking a new variety without a licence from the holder. However, plant breeders' rights do not prevent another breeder from using a variety in their own breeding programme.

Since genetic modification was developed, many patents have been filed covering the use of specific genes, types of genes, gene promoters, GM plant varieties and various technologies used to transform plants. Most of these patents have never been tested in court and it is still unclear to what extent they will be effective. Many people in the industry are sceptical about their usefulness but continue to file patents if only to ensure that they are not shut out by someone else doing so.

More confusion is generated by the fact that different countries have different patenting laws and interpretations. European patent examiners have been very stringent in making sure that biotechnology patents satisfy the three criteria listed above, while their American counterparts have been more inclined to accept a patent filing and let the courts sort it out later if the patent is challenged. Canada does not allow patents that cover plants or animals at all. The World Trade Organisation is attempting to uphold patents covering plant varieties worldwide under the General Agreement on Trade and Tariffs (GATT), but it has its work cut out.

The advantage for a company in holding a patent on the use of a particular gene in plant biotechnology is that the gene can then only be used under licence. Anyone who wants to incorporate the gene into their own breeding programmes, or even to grow a crop containing the gene, can only do so with permission from the company holding the patent. The biotechnology company has the right to charge a royalty for use of the gene or of crops that contain the gene. It is therefore able to make money out of its technology in ways that plant breeders' rights would not allow.

Patenting law does not allow for so-called 'bio-piracy', for example the raiding of a developing country's biological resources by a western company. A patent on a particular species of tree, or a traditional crop, would not satisfy the three criteria listed above and would be thrown out. However, it is possible to patent the idea of using a particular gene from any source to engineer a trait into a crop plant. Scientists in countries such as India and China are well aware of this and are keen to exploit the huge biological potential of their native flora, which dwarfs that available to scientists in temperate countries.

5.15. Loss of Genetic Diversity

It is widely believed that the introduction of GM crops will reduce biodiversity in agriculture, or more accurately it will reduce the genetic diversity available to plant breeders in the form of different crops and varieties. The great success of, for example, herbicide-tolerant soybean might at first sight appear to support this idea. However, the fact that a particular trait becomes popular with farmers does not necessarily mean that a single variety will be grown by everyone. The glyphosate-tolerant trait has been licensed by Monsanto to many other plant breeders, who have crossed the original line with their own varieties to produce new varieties that carry the herbicide-tolerance gene but are suited for particular local growing conditions. Approximately 150 different seed companies now offer glyphosate-tolerant varieties of soybeans.

Plant breeders are acutely aware of the risks of too much in-breeding and the potential of wild and unusual genotypes for providing genes that might improve a crop breeding line. For this reason, they conserve old and wild varieties in case they can be used in a breeding programme. Potato breeders, for example, have access to over 200 cultivated potato lines and 8,000 wild potato relatives that have been characterised. This is more genetic diversity than they are ever likely to know what to do with.

5.16. The Dominance of Multinational Companies

Much of the basic science and technology that underpins plant biotechnology was developed in the public sector in the USA, Europe and Asia, and plant biotechnology continues to attract considerable government funding for research all around the world, with China probably the heaviest investor. The commercial development of GM crops in the west, however, is dominated by six multinational companies: BASF, Bayer, Dow, DuPont, Monsanto and Syngenta. The domination of any industry by a small number of companies is always a concern, but it is hardly unique to plant biotechnology. Ironically, although the multinationals have retreated from Europe they will be able to return later if public opinion changes. The small and medium-sized homegrown companies and start-ups that might have competed with them, most of which were in the UK, did not survive.

Furthermore, the expense of developing new GM varieties, much of it associated with negotiating the regulatory system, means that only large companies have the resources to do it.

5.17. The StarLink and ProdiGene Affairs

StarLink was a trade name given to several GM maize varieties produced by Aventis that, like those varieties discussed in Section 3.6.2, had been engineered to be resistant to insects by the insertion of a *Cry* gene from *Bacillus thuringiensis*. However, unlike the successful GM maize varieties grown for human consumption, StarLink contained the *Cry9C* version of the gene instead of the *Cry1A* version. It also contained a gene imparting tolerance of the herbicide gluphosinate.

Problems arose with StarLink because, unlike the product of the *Cry1A* gene, that of the *Cry9C* gene does not break down easily in the human digestive system and is relatively heat resistant. For this reason, StarLink had never been approved for human consumption. However, in 1998 the EPA of the USA approved StarLink for commercial growing as an animal feed. It is not clear to what extent the other federal agencies involved in GM crop assessment, the FDA and APHIS were consulted on this decision. The EPA set a zero-tolerance level for the use of StarLink in human food.

Maize is an out-breeder, and some cross-pollination between StarLink and maize varieties destined for human consumption was inevitable. This should undoubtedly have been foreseen by the EPA and Aventis but apparently it was not. In mid-September 2000, *The Washington Post* reported that StarLink maize had been detected in processed maize snack foods already on the shelf in grocery stores. The products were recalled, the FDA became involved and in the end 300 maize products had to be taken off the shelf. Products were also recalled in Japan and Korea.

Aventis instructed its seed distributors in the USA to stop sales of StarLink seed corn for planting in 2001 and voluntarily cancelled its EPA registration for StarLink. However, Aventis predicted that it would take 4 years for StarLink to clear the food chain entirely. The company agreed to buy back the entire StarLink crop of the 2000 season at a premium price, something that must have cost them upwards of $100 million. The 1999 crop was already beyond recall. The incident left Aventis financially vulnerable, and it was subsequently acquired by Bayer.

Another incident involving the accidental mixing of a non-food GM crop with food intended for human use occurred in November 2002, although in this case the crop in question did not go beyond the farm. The biotechnology company involved was ProdiGene, a company that specialises in developing GM crops to produce pharmaceutical products (Section 3.17). Clearly, these crops are not meant to enter the food chain.

In 2001, ProdiGene contracted a Nebraska farmer to grow one of its experimental GM maize varieties. However, after the crop had been harvested the land was not treated any differently to land on which an ordinary crop had been grown. The next year the farmer planted soybean destined for human consumption on the same land. Inevitably, a small number of maize plants arising from spilled seed (referred to be farmers as 'volunteers') grew amongst the soybeans and a tiny amount (65 grams) of maize stalks were discovered in the harvested soybean seed. The FDA ordered the soybean crop, worth $2.7 million, to be destroyed.

Such a release in the UK would have to be monitored regularly during all stages of development. Post-harvest, the plants and/or seed would be stored in a secure area, all ventilation would be enclosed and all dust from extraction units or from drying and storage areas would be inactivated by autoclaving (heating under pressure). All straw and similar material would be returned to the release site and burned prior to cultivation into the soil. After harvest, the site would be ploughed and monitored. Irrigation would be used to encourage germination of volunteers and any re-growth or volunteers would be removed by spraying with an appropriate total herbicide. No crop would be sown on the site at least for the following year, probably the one after as well. The release site would be monitored for 2 years and any re-growth or volunteers would be removed by spraying with an appropriate herbicide or by cultivation.

All this would apply to a crop designed eventually for food use, never mind one making a pharmaceutical product. Clearly, similar practices should have been used in this case. However, this case does not change the fact that the use of GM plants to produce pharmaceutical products has enormous potential.

The StarLink and ProdiGene incidents should not have been allowed to happen, and they highlight the need to segregate crops designed for food and non-food uses. Regulatory authorities will have to decide whether out-breeding species such as maize are suitable plants for the purpose of making pharmaceutical and other non-food products. In fact, earlier in 2002 the United States Department of Agriculture ordered a different ProdiGene GM maize trial in Iowa to be destroyed to prevent cross-pollination with food maize growing nearby. Decisions such as that should be made before the crop is planted and some crops for which there is zero tolerance of presence in the food chain will have to be contained under glass.

5.18. The *Cauliflower Mosaic Virus* 35S RNA Gene Promoter

Another scare-story that has arisen in the GM debate in Europe is that the use of the *Cauliflower mosaic virus* 35S RNA gene promoter (usually called the CaMV35S promoter) (Section 2.11) in plant biotechnology represents some sort of human health risk. *Cauliflower mosaic virus* infects cauliflowers (hence its name) and other plants in the *Brassica* (cabbage) family. It does not infect animals, including humans, at all, but it infests most of the *Brassica* crop of the UK to some extent, and probably always has done. It is not, therefore, a good candidate for a food scare.

The CaMV35S gene promoter is only a small part of the viral genome. It is used in plant biotechnology where a gene introduced by genetic modification is required to be active everywhere in the plant. The supposed risk associated with it is that some sort of recombination could occur between this promoter and an animal virus, producing a new 'supervirus'. It is not clear exactly how this is supposed to happen, why it would be more likely to occur when the CaMV35S promoter is integrated into a plant genome as opposed to a viral genome, or why, even if it did occur, the resulting virus would be particularly dangerous.

5.19. Implications for Developing Countries

The effect that GM technology will have on developing countries is a question that has worried many people, and there has certainly been concern in developing countries either that they would become reliant on first-world companies for their seed or, conversely, that they would be denied access to GM technology altogether. These worries still persist.

It is important to note the potential benefits of GM technology for developing countries, as well as the problems. The potential of high vitamin A GM crops such as Golden Rice (Section 3.11), virus-resistant (Section 3.7) and high-yielding crops, and crops making vaccines (Section 3.17) to relieve famine, malnutrition and disease in developing countries is obvious. One of the more satisfying developments in the GM crop story of recent years is that developing countries are making their own decisions on the matter, much to the chagrin of western anti-GM pressure groups. Indeed, 18 of the 26 countries listed in Table 3.1 as GM crop producers are on the OECD's list of countries eligible to receive official development assistance (ODA), and three of them (Myanmar, Sudan and Bangladesh) are classed as Least Developed Countries as defined by the United Nations.

Developing countries are not all the same, of course; there is a big difference between countries such as India and China that have a large science base of their own and poorer countries that do not. Both India and China (defined as Lower Middle Income and Upper Middle Income Countries, respectively) are investing heavily in plant biotechnology, and China has developed several GM crop varieties for its own farmers (Table 5.1). In India, there was strong resistance to the cultivation of GM crops until the early 2000s. This resistance has broken down due to the success of insect-resistant (Bt) cotton, the first variety of which to be made available to farmers was introduced in 2002 (Section 3.6.1).

GM insect-resistant cotton has been grown in Burkina Faso, although not in 2016, and GM maize and cotton is grown extensively in South Africa. Elsewhere in Africa, arguably the continent with most need to exploit innovations in plant breeding, GM crops are still not being made available to farmers. An example of an African scientist who advocates

Table 5.1. GM crop varieties developed in China.

Species	Trait
Cotton	Insect resistant
Petunia	Colour altered
Tomato	Virus resistant
Tomato	Increased shelf-life
Sweet pepper	Virus resistant
Papaya	Virus-resistant

strongly that the potential of GM crops should at least be explored is Dr Florence Wambugu, the Founder Director and Chief Executive Officer of Africa Harvest Biotech Foundation International (AHBFI), or Africa Harvest. Wambugu, who studied for her PhD at the University of Bath in the UK, has stated that 'Having missed the Green Revolution, African countries know they cannot afford to pass up another opportunity to stimulate overall economic development through developing their agriculture'. However, African governments appear to be strongly influenced by opinion in Europe, at least on this issue.

An example of this was the controversy over the supply of food aid to Southern Africa in 2002. At first, encouraged by European activists, Malawi, Swaziland, Lesotho, Zimbabwe, Mozambique and Zambia all rejected food aid from the USA that contained GM maize, despite being threatened with famine as the result of a drought. As it became clear that people were going to die without the food aid, the activists became silent, fearing a backlash. The bogus health issue receded and the problem became one of future trade with Europe. The African governments feared that some of the imported seed would mix with seed being planted for future harvests and this would jeopardize trade with Europe once the drought was over. At first, the EU refused to reassure the Africans that GM crops were acceptable, afraid that this would threaten its trade bans on the importation of some GM foods from the USA. However, it eventually relented and gave the reassurances that were asked for. Most of the countries affected then agreed to accept the food aid, but Zambia, despite the dire need of its people for food, would not.

5.20. 'Terminator' Technology

'Terminator' technology is a term that has come to be used to describe the production of crop varieties that produce infertile seeds, or 'suicide seeds' as they are sometimes referred to. Pressure groups that oppose the use of genetic modification in plant breeding have long sought to associate GM crops with 'terminator' technology; indeed, the term 'terminator' and the prospect of biotechnology companies using such a technology to force farmers, particularly those in developing countries, to buy seed from them every year has been one of the most effective weapons in the anti-GM campaign. This is astounding given that none of the GM varieties in commercial use for food production produce infertile seed.

It is certainly possible to render the seeds of plants infertile using genetic modification; it can and has been done using non-GM methods as well. Similarly, both GM and non-GM methods can be used to render pollen infertile; indeed, it is an established technique in plant breeding for the production of hybrid seeds (Section 5.21), although the hybrid seed that is produced from such plants to be sold to farmers has its fertility fully restored. There may well be applications for the commercial use of sterile plants in specialist, small-scale applications such as the production of vaccines or other pharmaceuticals where it is important to prevent crossing and mixing with crops being grown for food. I am not aware of any breeder developing a variety, GM or otherwise, with sterile seeds for everyday use, and it is difficult to envisage why farmers would buy such a variety if it were launched.

5.21. Unintentional Releases

On 17th April 2000, Advanta Seeds told the UK government that GM-contaminated oilseed rape seed had been sold to farmers across the UK. The incident arose from inadvertent crossing of oilseed rape being grown for seed in Canada. The seed was to be a non-GM hybrid, and to ensure that all of the seed would be hybrid the seed-producing line was male-sterile, producing no pollen. Pollen was to be provided by the other would-be parental line which was planted alongside. This technique is used widely in plant breeding and relies on the pollinating parent producing pollen at exactly the right time to coincide with the seed-producing parent becoming receptive. Otherwise, in the absence of any pollen from the intended male donor, the seed-producing plants are pollinated by whatever compatible pollen is blowing in the wind. That is what happened in this case. These things happen in plant breeding but, on this occasion, some of the pollen in the wind was from a GM glyphosate-tolerant oilseed rape being grown on a nearby farm, and the seed was sold in Europe, specifically the UK, Sweden, France and Germany. The GM variety had not been approved for cultivation in Europe and an almighty hoo-ha ensued, despite the fact that only 1–3% of the seed was GM and the GM oilseed rape variety in question was grown widely in Canada (and still is), was approved for food use, and oil from it had been consumed by probably millions of people.

Several thousand hectares had been sown with the 'contaminated' seed in 1999, the seed had been harvested and processed, and the oil and meal consumed before the incident came to light; more was planted in 2000 and was in the field when the problem was recognised. However, not surprisingly, no ill effects were reported and there is no evidence of persistence of the GM trait in the UK or elsewhere in Europe.

Europe's strict laws governing the use of GM seeds and products make it almost inevitable that this sort of incident develops into a 'crisis'. Similar problems arise when crop products imported for use in food or feed are found to contain GM material that is not approved. In 2005, for example, it was reported that seed of an authorised Syngenta GM maize variety, Bt11, contained a small amount of seed from another transgenic line, Bt10. Bt10 contained the same Bt gene for insect resistance as Bt11, but the gene was expressed at a much lower level, making the protection against insect pests ineffective. Development of Bt10 had been stopped for this reason, but some Bt10 had persisted in Bt11 seed production for several years. The Bt11 variety was being grown widely in the USA and grain was being imported into Europe for food and feed use. The imported grain potentially contained small amounts of the unapproved variety, Bt10.

This should not have happened and American farmers who bought Bt11 seed would not have got the protection against insects that they paid for in a small proportion of their maize plants. However, there was no suggestion that the problem represented a threat to human health or the environment. Nevertheless, in 2005 the European Union required Member States to sample shipments of raw or minimally processed maize products coming from the USA at ports, feed mills and distributors' premises, at great expense. The UK's Food Standards Agency collected samples over a 4-week period in September and October of 2005 but was unable to detect any Bt10 grain.

5.22. Asynchronous Approvals

Another increasing difficulty for the European Union is the speed of approval and commercialisation of GM varieties outside Europe compared with the time it takes for new GM varieties to get through the European approval process. This leads to the problem of 'asynchronous approvals', with farmers outside Europe growing varieties that cannot yet be imported for food or feed use. Europe is not self-sufficient in food and feed production and, for example, has to import soybean from elsewhere for the production of animal feed. In 2009, the issue of asynchronous approvals led to USA farmers growing a new GM soybean variety that had not been approved in Europe to the extent that the European animal feed industry faced a shortfall in supply.

5.23. The UK Farm-Scale Evaluations

In 2000, the UK government was faced with the prospect of the first GM crops becoming available for commercial cultivation in the UK while the public remained fearful and hostile. In an attempt to instil confidence and to satisfy the demand that the potential effects of GM crops on the environment should be tested more thoroughly before they were grown widely, the government struck a deal with the biotechnology industry. This was called the **S**upply **C**hain **I**nitiative for **M**odified **A**gricultural **C**rops (SCIMAC) agreement. The industry undertook not to commercialise GM crops before 2003. In the meantime, farm-scale evaluations (field trials) would be carried out on the first generation of GM crops proposed for use in the UK in order to determine what effect they would have on the environment. The trials management committee included representatives from English Nature (the government agency responsible for nature conservation in England) and the Royal Society for the Protection of Birds.

The government stated that commercialisation after the farm-scale trials had been completed would not proceed if the GM crops were shown to have an adverse effect on biodiversity. This had never been asked before of an innovation in agriculture and is still not asked of new non-GM varieties or of new or old farming methods.

The GM crops in the trials were gluphosinate-tolerant (Liberty Link) oilseed rape and fodder maize from Bayer, and glyphosate-tolerant (Roundup-Ready) sugar and forage beet from Monsanto (Sections 3.5.1 and 3.5.2). These varieties were potentially close to commercialisation in the UK and the focus on them made sense from that point of view. However, it meant that the trials concerned only herbicide-tolerant crops used in conjunction with the appropriate herbicide, and did not enable a general assessment of GM crops or genetic modification to be made. Unfortunately, the trials also followed narrow crop management systems devised specifically for weed control. The results of the trials were announced in 2003 and showed that although biodiversity (that is the number and variety of weeds and insects) in the GM maize was higher than in the non-GM maize, for the other crops the GM varieties scored lower for biodiversity. Ironically, some maize farmers who took part in the trials were unimpressed with the weed control provided by the GM maize in combination with gluphosinate: farmers think their primary role is food production and weeds impact on their yield.

The UK media geared up for the release of the Farm-Scale Evaluation results with a steady stream of anti-GM headlines. After the report was published, many newspapers stated that the results represented 'the end of GM in the UK'. The UK government had put itself in a position where it would have been politically extremely difficult to allow commercialisation of the GM crops that had a bad biodiversity report in the trials. However, it was even unwilling to allow commercialisation of Bayer's GM maize variety, stating that it would consider the issue of segregation and co-existence with non-GM crops first and giving no deadline for a decision. Bayer then announced that they would not proceed with commercialisation.

The UK Farm-Scale Evaluations programme provided a lot of information on the insects and weeds in UK farmland. They did very little to inform the debate on GM crop use in the UK. The scientists involved did a very professional job in the execution of the experiment and the reporting of the data, but failed to ensure that the media were aware of the limitations of the study. Indeed, they were hopelessly naïve in expecting the results of the study to be reported in a balanced way without being much more proactive in their dealings with the media. There is no doubt that the process put any prospect of GM crop cultivation in the UK back by years.

5.24. Conclusions

Genetic modification is now an established technique in plant breeding in many parts of the world. While not being a panacea, or replacing every other method in plant breeding, it does enable plant breeders to introduce some traits into crop plants that would not be possible by other methods. GM crops now represent almost 10% of world agriculture by area, and are being used in developed and developing countries. Farmers who use them report one or more of greater convenience, greater flexibility, simpler crop rotation, reduced spending on agrochemicals, greater yields or higher prices and increased profitability at the farm gate as the benefits.

That does not mean that every farmer who has the option to use GM varieties does so. GM varieties enter a very competitive marketplace, and suppliers offering alternatives reduce their prices or offer other incentives to make their own products competitive. Nor is it the purpose of this book to tell farmers what varieties, GM or otherwise, they should or should not grow, only that UK and European farmers should be given the choice of growing GM crops that have come through the testing and regulatory processes. Otherwise they will find it increasingly difficult to compete with overseas farmers.

The delay in allowing plant biotechnology to develop in Europe has already damaged the European plant biotechnology industry significantly and is putting European agriculture at an increasing competitive disadvantage. The consequences of losing the debate, which is where we are at the time of writing, can be summarised as:

- Currently no GM crops are being developed specifically for European conditions or the European market.
- The agricultural biotechnology industry is currently focused on obtaining consent for import of more GM crop products into Europe, rather than for permission to market GM crop seed for cultivation in Europe.
- European farmers are increasingly disadvantaged in a competitive global market, competing with GM crops but unable to use them.
- The agricultural biotechnology multinationals have closed or down-sized their biotechnology research and development operations in Europe.
- The small agricultural biotechnology start-up industries that appeared in Europe in the 1980s and 1990s have been lost.
- The USA, China, India, Brazil, Argentina and others have a huge lead over Europe in a key 21st-Century technology.
- While new techniques for genome editing may enable Europe to get over the GM impasse, perhaps by leap-frogging it altogether, industry investment in genome editing will be minimal while the European Commission refuses to give any information on whether and how genome edited crops will be regulated.

The GM crop issue first came to the attention of the UK public in 1996, already more than two decades ago. The UK appears to be no closer to commercial GM crop cultivation now than it was then. However, the world has changed in that time. The 1990s were in the era of crop surpluses in Europe, with plentiful cheap food and still relatively cheap energy. With news stories of European farmers being paid to destroy produce being a regular feature, increasing crop yield was not even seen as a worthwhile target for plant scientists. Food shortages elsewhere in the world were talked of as issues of distribution rather than production. Warnings that a small surplus of food was much better than too little, and that the situation would not last, were ignored.

The issue of crop yield and the need to increase food production crashed back on to the global agenda in the second half of the last decade as food prices soared. The price of wheat on the London International Financial Futures and Options Exchange (LIFFE), for example, rose to a record £198 in 2008 as a result of increasing demand and severe drought in Australia. The Australian drought could have been seen as a one-off event, but there was concern that it was the sort of extreme weather event that would occur with increasing regularity through the 21st Century as a result of climate change. Increasing demand was also something that could not be ignored. World population is 7.6 billion at the time of writing, and the United Nations estimates it will increase to 11.2 billion by the year 2100. In addition, greater prosperity in highly populated, economically rapidly emerging countries such as China and India is leading understandably to a demand for a better diet, in particular a demand and ability to pay for more meat, much of it from animals fed at least in part on grain. On top of that, crops are now being used for fuel as well as food (Section 3.10), with more than a third of the USA's maize crop, for example, now being used for bioethanol production. All this means that the demand for crop products is increasing faster than the population is growing. In China, for example, while population doubled between 1962 and 2012, wheat consumption increased six-fold, from just under 20 to 123 mt, with *per capita* consumption trebling from 29 to 92 kg.

Farmers responded to the increase in food prices in 2007–2008 by increasing production, as they always have done, and prices fell back. However, 2010 saw another major commodity crop exporter, Russia, suffering from drought, and prices climbed again. In December 2010, the LIFFE price hit a new record of £200 per tonne. Prices will no doubt continue to fluctuate, but arguably the era of cheap food and global food surpluses has ended.

Agriculture will face several difficult challenges in the coming decades: as well as population growth and climate change, there will be problems of fresh water supply, peak oil (the point when the maximum rate of global petroleum extraction is reached, after which the rate of production enters terminal decline), competition for land use, the need for wildlife conservation, soil erosion, salination and pollution. Furthermore, while agriculture will be affected by climate change, it will also be expected to contribute to efforts to cut greenhouse gas emissions in order to avoid the worst-case scenario predictions of global temperature increases. In the UK, for example, the Climate Change Act (2008) commits the country to an 80% reduction in greenhouse gas emissions by 2050 across all sectors of the economy, and farming is responsible for about 7.4% of total UK emissions. If agriculture is going to meet these challenges, plant breeding will have to play its part and plant breeders will have to be able to use every tool that is available, including biotechnology.

Index